POLARONS

Emin provides experimental and theoretical graduate students and researchers with a distinctive introduction to the principles governing polaron science. The fundamental physics is emphasized and mathematical formalism is avoided. The book gives a clear guide to how different types of polaron form and the measurements used to identify them. Analyses of four diverse physical problems illustrate polaron effects producing dramatic physical phenomena.

The first part of the book describes the principles governing polaron and bipolaron formation in different classes of materials. The second part emphasizes the distinguishing electronic-transport and optical phenomena through which polarons manifest themselves. The book concludes by extending polaron concepts to address critical aspects of four multifaceted electronic and atomic problems: Large-bipolaron superconductivity, electronic switching of small-polaron semiconductors, electronically stimulated atomic desorption, and diffusion of light interstitial atoms.

DAVID EMIN retired from Sandia National Laboratories and is currently Adjunct Professor in the Department of Physics and Astronomy, University of New Mexico. He is best known for his work on polaron formation and motion and is often cited for his seminal contributions to the theories of self-trapping and hopping conduction's phonon-assisted transition rates, Hall effect and Seebeck effect.

POLARONS

DAVID EMIN

University of New Mexico

CAMBRIDGE
UNIVERSITY PRESS

CAMBRIDGE
UNIVERSITY PRESS

University Printing House, Cambridge CB2 8BS, United Kingdom

One Liberty Plaza, 20th Floor, New York, NY 10006, USA

477 Williamstown Road, Port Melbourne, VIC 3207, Australia

314-321, 3rd Floor, Plot 3, Splendor Forum, Jasola District Centre, New Delhi - 110025, India

79 Anson Road, #06-04/06, Singapore 079906

Cambridge University Press is part of the University of Cambridge.

It furthers the University's mission by disseminating knowledge in the pursuit of education, learning and research at the highest international levels of excellence.

www.cambridge.org
Information on this title: www.cambridge.org/9780521519069

First published 2013

A catalogue record for this publication is available from the British Library

Library of Congress Cataloging in Publication data
Emin, D. (David)
Polarons / David Emin.
pages cm
ISBN 978-0-521-51906-9 (hardback)
1. Polarons. 2. Electrons. I. Title.
QC176.8.P62E45 2013
539.7'21–dc23
2012022011

ISBN 978-0-521-51906-9 Hardback

Contents

Preface

In 1927 Born and Oppenheimer formalized the notion that electrons in condensed matter generally follow the relatively slow motions of atoms which themselves adjust to the electrons' averaged locations. In 1933 Landau advanced this idea in suggesting that an electron added to an insulator such as sodium chloride could become *self-trapped*. The electron is then bound in the potential well produced by displacing the equilibrium positions of surrounding atoms from their carrier-free values to those stabilized by the carrier's very presence. Self-trapped carriers move very slowly since they only move when the associated atoms move.

The term *polaron* often refers to the unit comprising a self-trapped carrier and the associated pattern of displaced atomic equilibrium positions. However, the term polaron is also used in a more general sense to designate an electronic carrier and the altered atomic motions induced by it. Then polarons whose electronic carriers are not self-trapped are labeled as *weak-coupling polarons*. By contrast, polarons whose electronic carriers are self-trapped are termed *strong-coupling polarons*.

In some instances polaron motion can be regarded as quasi-free albeit with occasional interruptions by scattering events. The effective mass of a weak-coupling polaron is only slightly larger than that of its bare electronic carrier. By contrast, a strong-coupling polaron's effective mass is very much larger than that of the bare electronic carrier.

These scattering treatments of polaron motion presume its coherence. The motion's transfer energy must then exceed the energy variations and uncertainties which accompany the polaron's movement and scattering. For instance, a strong-coupling polaron's motion is coherent if the electronic transfer energy between adjacent sites exceeds changes of its self-trapped carrier's energy which occur as atoms move classically to enable this transfer. This condition is satisfied when the electronic transfer energy is large enough for the self-trapped carrier to extend over at least several sites. Thus a *large*-polaron's motion is coherent.

By contrast, a *small* polaron has its self-trapped carrier primarily confined to a single site. A stable small polaron only forms when its self-trapped carrier's electronic transfer energy is sufficiently small. For such small values of the electronic transfer energy the self-trapped carrier's response to classical atomic motion is incoherent. Small-polaron transport then proceeds by a succession of thermally assisted jumps.

In an unperturbed ideal crystal small-polaron motion becomes coherent at low temperatures when atoms' motions become non-classical. Then the inter-site motion of a self-trapped carrier occurs as the atoms of the small-polaron's atomic displacement pattern collectively tunnel between geometrically equivalent displaced equilibrium positions. However, the small-polaron transfer energy associated with such motion is extremely small, much smaller than the characteristic phonon energy. Small-polaron transfer energies are usually small enough to be dwarfed by the site-to-site energy differences of a real material. As a result small-polaron motion in actual materials is generally regarded as incoherent even at very low temperatures.

This physical picture was already established half a century ago when I began my studies of the formation and properties of polarons. Nonetheless, important issues remained unresolved. It was necessary to proceed beyond oversimplified models to understand how polarons and self-trapped paired carriers' bipolarons form. Hopping transport needed to be treated beyond the unrealistic non-adiabatic limit, generally restricted to electronic transfer energies that are much smaller than even phonon energies. Means were needed to distinguish between intrinsic small-polaron hopping and extrinsic multi-phonon-assisted hopping involving defects and dopants. Many effects of the shifts of atoms' vibration frequencies arising from self-trapped carriers' polarizabilities required consideration. These topics are addressed within this book's discussion of polarons.

The book's seventeen chapters are divided into three sections. In the first section I explain how and when different types of polaron form. In the second section I address polarons' distinctive electrical transport and optical properties. It is through these properties that polarons can be identified. The final section considers four dissimilar electronic and atomic phenomena for which polaronic effects may be critical.

Throughout this discussion I stress the underlying physics governing polaron formation and motion. I eschew formal mathematics and oversimplified models. I make no attempt to review the rich formalisms that have been developed to address the many-body problem of interacting electrons and phonons. As such, this book does not mirror the totality of present polaron research.

This exposition on polaron physics is an idiosyncratic account. I emphasize several important consequences of carrier-induced softening, a frequently neglected phenomenon. I also address some topics such as small-polarons' hopping conduction and the Hall effect in exceptional detail in order to highlight overlooked features of these phenomena. Furthermore, I even advance the unorthodox view that some heavily doped materials' electronic hopping conduction occurs between defect-related sites rather than between intrinsic sites. I also argue that a phonon-mediated attraction between large bipolarons can drive their condensation into a liquid. This large-bipolaron liquid, analogous to liquid ^{4}He, could be a precursor to further condensation into a superconducting state. I present a heterodox explanation of the driving mechanism of electronic switching in low-mobility solids. Finally, the concepts of small-polaron formation and motion are extended to address electronically stimulated atomic desorption and light-atom diffusion.

To put this treatise's distinctive ideas in context I now succinctly describe them within a summary of the polaron physics' intellectual framework. These concepts involve (1) polaron and bipolaron formation, (2) polaron and bipolaron properties and (3) several problems to which extensions of core polaron notions have been applied.

Polaron formation concerns an electronic carrier interacting with relatively slow-moving atoms in accord with the adiabatic principle. Atoms accommodate to a suitable average of the carrier's position while the carrier adjusts to atoms' instantaneous positions. *Carrier-induced softening* describes the energy lowering resulting from carriers adjusting to atoms' vibrations. The binding energy of a weak-coupling polaron arises solely from the dynamics of this electronic response. As such, the energy lowering associated with forming a weak-coupling polaron vanishes in the adiabatic limit. In this limit atoms' vibration frequencies approach zero as atomic masses become infinite while vibrations' stiffness constants remain finite. By contrast, a strong-coupling polaron involves atoms vibrating about the equilibrium positions they assume to form the potential well within which the self-trapped carrier is bound. Thus, a strong-coupling polaron's energy lowering does not vanish in the adiabatic limit where atoms' vibrations are suppressed.

The size of a self-trapped state generally depends on its carrier's electron–phonon interactions. A carrier has long-range electrostatic interactions with the displaceable ions of a polar medium. A carrier also has short-range interactions with the displaceable atoms that it contacts. These electron–phonon interactions drive carrier-induced shifts of atoms' equilibrium positions and vibration frequencies. Both effects alter the coupled system's free energy. Nonetheless, the size of a carrier's self-trapped state is usually determined by minimizing the energy of its strong-coupling polaron in its adiabatic limit where atoms' vibrations are ignored.

With just the three-dimensional long-range electron–phonon interaction the spatial extent of a self-trapped state is a continuous function of the physical parameters. Then the self-trapped carrier's radius is $R_\mathrm{p} \equiv (\hbar^2/m^*e^2)[\varepsilon_\infty/(1 - \varepsilon_\infty/\varepsilon_0)]$, where m^*, ε_0 and ε_∞

denote the electronic carrier's effective mass and the material's static and optical dielectric constants, respectively. This radius is usually a few angstroms: (1) $\hbar^2/m^*e^2$ is just the Bohr radius (≈ 0.5 Å) when m^* is set equal to the free-electron mass and (2) $\varepsilon_0 \sim 5$–10 and $\varepsilon_\infty \sim 2$–$3$ for typical ionic solids. Thus the polaron produced with just the long-range electron–phonon interaction is generally a large polaron.

By contrast, a multi-dimensional self-trapped state formed with just the short-range electron–phonon interaction is a discontinuous function of its physical parameters. Such self-trapping is dichotomous: the carrier either self-traps with the smallest acceptable spatial extent thereby forming a small polaron or it does not self-trap at all.

Large- and small-polaron states can coexist separately when these electron–phonon interactions are combined. Alternatively combining these interactions can induce a large polaron to *collapse* into a small polaron. The confining effect of defects and the localizing tendency of disorder can also induce carriers to collapse into severely localized small-polaronic states.

A bipolaron forms when two carriers find it energetically favorable to share a common self-trapping potential well. Forming a bound state with just the long-range electron–phonon interaction requires an especially large ratio of static-to-optical dielectric constants. Even then, without correlating the positions of these two carriers to reduce their mutual Coulomb repulsion, the long-range electron–phonon interaction is unable to stabilize the self-trapping of paired carriers with respect to two separated polarons. However, the addition of the short-range electron–phonon interaction can lower the pair's energy enough to stabilize a large bipolaron. These large bipolarons are slightly smaller than a large polaron. If, however, the short-range electron–phonon interaction is too strong the bipolaron will collapse into a severely localized small bipolaron. The domain for large-bipolaron formation increases with the ratio of static-to-optical dielectric constants. In other words, large-bipolaron formation requires readily displaceable ions.

Degeneracy of carriers' electronic energy levels enhances the contribution of carrier-induced softening to bipolaron formation. A *softening bipolaron* is a singlet bipolaron whose formation is driven primarily by carrier-induced lowering of its vibrations' free energy. The four-fold orbital degeneracy of the frontier orbitals of boron carbides' icosahedra makes them prime candidates for the formation of softening bipolarons. Softening-bipolaron formation is suggested by the high density of holes which hop between icosahedra without producing a commensurate paramagnetic susceptibility, ESR or polaronic absorption but generate a large softening-type enhancement of boron carbides' Seebeck coefficients.

The very slow coherent motion of a massive large polaron is disrupted by indigenous phonons which scatter as they impinge on the region softened by the polarizable self-trapped carrier. A large polaron's hefty momentum ensures that it will be only weakly scattered by phonons whose moderate dispersion imbues them with relatively small momenta. This weak scattering enables a large polaron to still move with a moderate mobility. A large polaron's scattering by phonons causes its mobility to fall with rising temperature as the density of these phonons increases. Concomitantly the polarons' strong scattering of indigenous phonons reduces their thermal conductivity.

Small polarons move by thermally assisted hopping with very low mobilities that increase as the temperature is raised. A small-polaron-mobility's ever steepening rise with increasing temperature culminates with Arrhenius behavior when the temperature is raised above the characteristic phonon temperature.

The Arrhenius region's activation energy generally increases as the jump distance increases. The specifics of this effect depend on (1) the range of the predominant electron–phonon interaction, (2) the spatial extent of the self-trapped carrier, and (3) pertinent phonons' dispersion. By itself, this effect, ignored by oversimplified models, promotes small polarons making short hops. Polaron hops are also restricted to nearby sites by steric constraints and by their relatively large electronic transfer energies.

Small-polarons' localized electronic states often densely fill space insuring that hopping distances are just small multiples of states' radii. The resulting electronic transfer energies are usually larger than the energies of phonons with which carriers interact. Electronic transfer is then *adiabatic*, fast enough for a carrier to transfer whenever atoms' motions afford it the opportunity. Polaron hops are often adiabatic. By contrast, *non-adiabatic* hopping occurs when electronic transfer energies are so much less than phonon energies that a carrier only infrequently transfers when atomic motion affords it the opportunity. Such small transfer energies can occur when hopping lengths greatly exceed localized states' radii. Low-temperature hopping among dopants of a very lightly doped semiconductor is generally regarded as non-adiabatic.

An adiabatic semiclassical polaron hop occurs as a carrier sloshes between the sites of a double well as it changes in response to atoms' motion. This jump-related extended motion lowers vibrations' frequencies and thereby enhances their entropy. The enhancement of vibrations' entropy is nearly proportional to the jump rate's activation energy. This relationship characterizes the *Meyer–Neldel compensation effect*. That is, the jump rate's temperature-independent pre-factor is enhanced by an entropic factor which rises exponentially with the rate's activation energy. Thus the reduction of a semiclassical adiabatic jump rate's thermally activated factor tends to be compensated by an increase of the rate's temperature-independent factor.

A carrier's hopping is usually viewed as a succession of temporally uncorrelated jumps. This picture emerges from treating hops in the non-adiabatic limit where transfers are arbitrarily rare. However, adiabatic transfers occur faster than hops' atomic displacements form or relax. Thus a carrier's adiabatic hopping occurs in flurries. It consists of bursts of hopping activity compensated by relatively long dormant periods.

An electric current is driven as an applied electric field promotes carriers' hopping in concert with it. These hops tend to increase in length when the applied field is strong enough to overcome the rise of the hopping activation energy with jump distance. In ionic and polar materials this competition can cause the mobility's high-field activation energy to fall in proportion to the square root of the electric field thereby generating *Frenkel–Poole behavior.*

Small bipolarons form when electronic carriers self-trap as severely localized singlet pairs. However, the hopping of small bipolarons is not simply analogous to that of small polarons. Rather, small-bipolarons' carriers find it energetically favorable to break apart as they hop. This pair-breaking manifests itself in small-bipolarons' electronic transport and magnetic properties.

A polaron or bipolaron's Seebeck coefficient, the entropy transported per carrier divided by its charge, is affected by its carriers' interactions. Most generally, the change in configuration entropy arising from the addition of a charge carrier is enhanced by the narrowness of strong-coupling polaron bands. In some instances, the increase of vibrations' entropy produced by carrier-induced softening may also significantly enhance Seebeck coefficients.

The Hall mobility addresses the deflection of a moving carrier by a magnetic field. A conventional carrier's Hall mobility differs little from its drift mobility: its trap-free drift velocity divided by the electric field that drives its flow. However, the Hall and drift mobilities for a small polaron differ from one another in magnitude and temperature dependence. Moreover, the sign of the small-polaron Hall effect is often anomalous. That is, the small polaron is deflected in an opposite sense to that of a conventional carrier possessing a charge of that sign. The sign of a small-polaron's Hall effect depends on the symmetry of the self-trapped carrier's local orbital and the geometrical arrangement of sites between which it hops. Thus, Hall mobility measurements can help distinguish between a small-polaron's intrinsic hopping and a carrier's hopping between defects, impurities or dopants.

Finally, I reiterate the distinctive ideas applied to the four topics presented in the third section of this book. These subjects are (1) bipolaronic superconductivity, (2) electronic threshold switching in low-mobility semiconductors, (3) electronically stimulated desorption from surfaces, and (4) light atoms' hopping diffusion.

Similarities between the superfluidity of liquid ^{4}He and the superconductivity of electronic carriers have fueled the long-standing speculation that superconductivity can result from local singlet electron pairs which behave as mobile bosons. Here the mobile pairs are identified with large bipolarons. Large bipolarons can form within especially polarizable ionic and polar media. The electronic polarizability of large bipolarons' self-trapped carriers promotes their softening of the vibrations to which they are coupled. The collective effect of this softening is to produce a mutual phonon-assisted attraction between large bipolarons. This mutual attraction facilitates large-bipolarons' condensation into a liquid analogous to that formed by ^{4}He. As in BCS superconductivity the stabilization of the condensed state increases with participating atoms' vibration frequencies. This scenario is suggested as an avenue to the realization of bipolaronic superconductivity.

Driving sufficiently large currents through numerous crystalline and non-crystalline semiconductors in which small-polaron formation is suspected causes them to switch from states of low conductance to states of much higher conductance. Materials having very low small-polaron-like mobilities often concomitantly possess very large carrier densities. These carrier densities will be driven still higher near interfaces with leads having higher mobility carriers as increasingly large currents are driven through them. Switching occurs when the carrier density becomes large enough to destabilize small-polaron formation with respect to electronic carriers remaining free. This scenario describes a general mechanism for electronic threshold switching.

The theory of self-trapping has been extended to address how the introduction of a delocalized electronic excitation on a surface can lead to an atom's desorption. The surface excitation's self-trapping localizes it on a surface bond, breaking the bond and releasing its atom. Depending on the magnitude of the surface excitation's kinetic energy, the process will be spontaneous or delayed by the necessity of transcending a barrier to the excitation's self-trapping.

Finally, the hopping diffusion of light atoms is treated as a generalization of the hopping of a small polaron. The physical significance of the high-temperature Arrhenius behavior's activation generally differs from that of a small polaron. The dependences of the diffusion constant's activation energy and its pre-exponential factor on a light-atom's isotopic mass provides a means of distinguishing the generalized treatment from literal application of the theory developed to describe the hopping of a self-trapped electron.

Acknowledgments

I was introduced to polarons half a century ago as a graduate student of Ted Holstein. Over time Ted and I became colleagues and close friends.

When I began to work with Ted we primarily viewed the study of polarons as an interesting academic pursuit. Many people in the condensed-matter community were unfamiliar with the concept of self-trapping and had never heard the term polaron. The idea of self-trapping received little or no mention in most introductions to solid-state physics.

It was uncertain that strong-coupling polarons were more than an oddity for which there were but a few examples. Over time reports of polaron effects grew. Evidence of polarons was reported in some transition-metal oxides, molecular solids and non-crystalline semiconductors. Experimentalists were bringing polarons to life.

Our simple models did not provide a satisfactory explanation of why these charge carriers formed polarons. It now seems that combinations of long-range and short-range electron–phonon interactions with the localizing effects of defects, impurities and disorder often trigger polaron formation.

This book offers an overview of what I have learned about polarons. Learning doesn't occur in a vacuum. I am grateful to the many people who have helped me.

I thank my collaborators for sharing work on a wide range of interesting problems. I am especially indebted to my experimental colleagues and collaborators for the data they present, the diversity of intricate materials they investigate and the questions that they and their work raise.

I was also lucky to have spent most of my career associated with Sandia National Laboratories. From my arrival in 1969 through my formal retirement I was continually able to devote myself fully to studying fundamental condensed-matter physics. My managers always supported my efforts with a sense of *noblesse oblige*. Science had not yet grown into a big enterprise.

I also appreciate the hospitality shown me by the Department of Physics and Astronomy of the University of New Mexico over the last 30 years. During this time I have collaborated with its faculty as well as its post-doctoral and graduate students. UNM has become my primary habitat since my retirement from Sandia.

For half a century I have received unflagging encouragement and support from my wife Shirley. She is remarkable in her continuing interest in learning about all of the science I study. I cannot imagine how my research and the writing of this book would have progressed without her steadfast and sensitive devotion.

Finally I wish to thank the editorial staff of Cambridge University Press for their continuous assistance, patience and understanding. It has taken quite a few years for the idea for this book to come to fruition.

1

Succinct overview

The genesis of polarons is often ascribed to a short paper published by Landau (Landau, 1933). That paper considers an electronic charge carrier moving slowly through an ionic solid. The carrier lingers long enough in a locale for the surrounding ions to assume shifted equilibrium positions as illustrated in Fig. 1.1. Altering the equilibrium positions of the ions establishes a potential well for the carrier. This potential well may be sufficient to bind the carrier. If the carrier's binding energy is greater than the characteristic phonon energy the bound carrier will see its potential well as nearly static. The carrier is then *self-trapped*, bound in the potential well its very presence induces. The carrier cannot escape without significant movement of the ions.

The self-trapped carrier together with the displaced ions has been defined as a strong-coupling *polaron*. The term polaron was coined in cognizance of the strong forces between a carrier and the ions of a polar material. However, the argument for self-trapping is general. As such, self-trapped carriers are reported among all classes of condensed matter. Polaron formation is energetically stable when the self-trapped carrier's binding energy exceeds the strain energy expended in displacing surrounding nuclei so as to produce the potential well that binds the carrier.

Even when the forces between the carrier and ions are too weak to produce self-trapping the carrier and the surrounding ions affect one another. The composite quasi-particle, comprising the carrier and surrounding ions whose motions are altered by the carrier's presence, is often termed a weak-coupling polaron.

Studies of self-trapping generally employ the adiabatic approach. The adiabatic method presumes that an electronic carrier adjusts to atoms' positions while also modifying atomic motion. Electronic carriers (1) shift atoms' equilibrium positions and (2) alter the stiffness constants which govern atoms' harmonic motions.

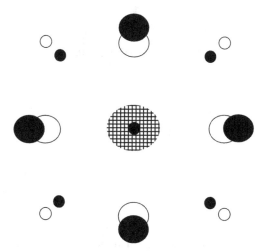

Fig. 1.1 An excess electron (hatched area) held on a cation shifts the equilibrium positions of the surrounding anions (filled large circles) and cations (filled small circles) from their carrier-free positions (open large and small circles). These shifts of ions' equilibrium positions produce a potential well for that carrier.

Strong-coupling polarons primarily shift atoms' equilibrium positions whereas weak-coupling polarons modulate atoms' vibration frequencies.

Adiabatic studies of self-trapping explicitly demonstrate that self-trapping is a nonlinear phenomenon. In particular, the self-trapping potential depends on the wavefunction of the self-trapped carrier. As a result, the nature of the self-trapped state depends critically on the details of the interaction between a carrier and the surrounding medium. For example, the spatial extent of a self-trapped state varies continuously with material parameters when the carrier interacts with the ions of a polar medium solely via the long-range Coulomb interaction. By contrast, the self-trapped state collapses to the smallest possible physical unit when the interaction between a carrier and the atoms is only via the short-range interaction characteristic of a covalent solid.

Self-trapping becomes more intricate in real solids for which a carrier's interactions with its surroundings have both long-range and short-range components. Molecular solids permit several types of molecular polaron. All told, self-trapping is sensitive to a material's anisotropy and its disorder. Localizing, constraining or slowing a carrier all foster its self-trapping.

A strong-coupling polaron is identified by its distinctive properties. A strong-coupling polaron possesses an absorption produced by optically exciting its self-trapped carrier. These phonon-broadened absorption bands are akin to those of trapped carriers. Since a self-trapped carrier can move only when the involved atoms move, a strong-coupling polaron must move very slowly. Concomitantly, its energy band is narrower than the characteristic phonon energy and its effective mass is correspondingly large.

Strong-coupling polarons manifest two distinct types of transport properties. A *large polaron* extends over several structural units and thereby moves with a coherence length exceeding the characteristic inter-site separation. The very heavy mass of such a polaron can lead to its being only weakly scattered by phonons. This weak scattering can compensate its large mass to yield a mobility ($\gg 1$ cm^2/V-sec) that falls with increasing temperature.

A *small polaron* is confined on a single structural unit and typically moves incoherently. A carrier's motion is then usually described as occurring via a succession of occasional phonon-assisted jumps between adjacent sites. The small-polaron hopping mobility is generally very low, $\ll 1$ cm^2/V-sec, and rises with increasing temperature. A distinguishing feature of such incoherent motion is the occurrence of Hall-effect sign anomalies. In these instances a small polaron is deflected by a magnetic field in the opposite direction than a classical particle of the same sign. Figure 1.2 summarizes the relationships between different types of polaron along with their principal characteristics.

Polarons interact with one another through several effects. Same-signed self-trapped carriers repel one another through their mutual Coulomb repulsion. Polarons also interact with one another through the interference of their patterns of atoms' shifted equilibrium positions and vibration frequencies. These interference effects can result in attractive interactions through which like-signed polarons may find it energetically favorable to pair as *bipolarons*.

Interactions among polarons and bipolarons also foster their collective behavior. Large polarons and bipolarons may be able to flow collectively as a liquid. Alternatively, polarons and bipolarons can order in a manner that is commensurate with the underlying lattice structure resulting in their being pinned and unable to flow. In addition, increasing the density of carriers tends to diminish their ability to self-trap as they compete to displace the same atoms. Driving the carrier density high enough can thereby destabilize carriers' self-trapping. Switching between different situations underlies various polaron-related phase transitions. These phases can be electrical insulators, small-polaron semiconductors, metals and even perhaps superconductors.

Measurements of optical, magnetic and electronic transport properties indicate polarons in a wide range of condensed matter. These materials include (1) ionic solids such as KCl, (2) magnetic insulators such as MnO, (3) molecular solids such as S_8, (4) many glasses (e.g. transition-metal-oxide glasses) and (5) liquids such as water in which electronic carriers are *solvated*. This book will describe how different types of polaron form and the properties through which they can be identified.

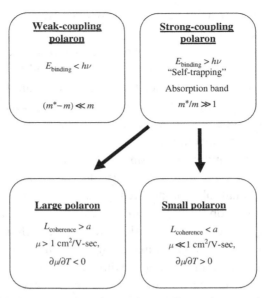

Fig. 1.2 Essential characteristics of a weak-coupling polaron and the two types of strong-coupling polaron, a large polaron and a small polaron, are summarized. A self-trapped carrier is bound with energy $E_{binding}$ in the potential well produced by shifts of surrounding atoms' equilibrium positions. Self-trapping occurs when the bound carrier's electronic frequency $E_{binding}/h$ exceeds these atoms' characteristic vibration frequency v. Optical excitation of the self-trapped carrier produces an absorption band. The mass associated with an elementary translation of the unit comprising the self-trapped carrier and its pattern of displaced atomic equilibrium positions m^* usually greatly exceeds the mass of an unbound carrier m. The polaron is deemed large when the spatial extent of its self-trapped carrier R_p exceeds the lattice constant a. A large polaron generally moves coherently with a moderate mobility μ which falls with increasing temperature. By contrast, a polaron is regarded as small when it collapses to a single structural unit. A small polaron usually moves incoherently via phonon-assisted hopping with a very small mobility which increases with increasing temperature.

The polaron problem has been extended to apply to other situations in which an agile entity interacts with its relatively sluggish environment. For instance, the polaron approach has been employed to address the interaction of an exciton with surrounding atoms. Light interstitial atoms (e.g. hydrogen and its isotopes) diffusing among the relatively heavy atoms of a metal has been treated as a generalization of small-polaron hopping. The concept of a magnetic polaron arises from considering the interaction of a carrier with the localized magnetic moments of a magnetic insulator. Moreover, a carrier in a magnetic solid also generally interacts with its atomic displacements. Synergy between the effects of a carrier's interactions with magnetic moments and atomic displacements is evident as misalignment of local moments triggers the collapse of a carrier into a small polaron. Magnetic and electronic transitions are thereby linked. These extensions of the polaron problem are also discussed here.

Part I

Polaron Formation

2

Electron–phonon interactions

Electron–phonon interactions describe how the energy of an electronic charge is affected by alterations of the positions of the atoms of the encompassing medium. Polaron problems usually take a charge carrier's energy to depend linearly on shifts of nuclear positions. Here, for simplicity, the medium is represented as a continuum. Then the net effect of deforming the continuum by $\Delta(\boldsymbol{u})$ at positions designated by $\boldsymbol{u}$ is to shift the effective potential energy of a carrier at $\boldsymbol{r}$ by

$$V(\boldsymbol{r}) = \int d\boldsymbol{u} Z(\boldsymbol{r} - \boldsymbol{u})\Delta(\boldsymbol{u}), \tag{2.1}$$

where $Z(\boldsymbol{r} - \boldsymbol{u})$ defines the magnitude and range of the electron–phonon interaction.

2.1 Long-range electron–phonon interaction

Ionic and polar materials support a long-range component of the electron–phonon interaction. Pairing anions with cations enables the Coulomb potential of these materials to be represented as that from an array of electric dipoles. Altering the separation between the ions of a dipole shifts the potential seen by a carrier. The modulation of the carrier's potential energy at $\boldsymbol{r}$ due to altering the charge separation of a dipole centered at $\boldsymbol{u}$ is governed by

$$Z_{\mathrm{LR}}(\boldsymbol{r} - \boldsymbol{u}) \equiv - \frac{\sqrt{\dfrac{e^2}{4\pi}\left(\dfrac{1}{\varepsilon_\infty} - \dfrac{1}{\varepsilon_0}\right)\dfrac{k}{V_{\mathrm{c}}}}}{|\boldsymbol{r} - \boldsymbol{u}|^2} \cos\theta, \tag{2.2}$$

where (as illustrated in Fig. 2.1) θ is the angle between the electric dipole and the vector $\boldsymbol{r} - \boldsymbol{u}$. The constants in Eq. (2.2) are the carrier's charge e, the material's

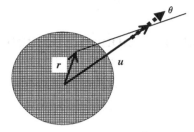

Fig. 2.1 An electric dipole (dotted arrow) directed along the vector from the carrier's centroid **u** interacts with its charge (hatched area) at **r**.

optical and static dielectric constants ε_∞ and ε_0, the Hooke's law stiffness constant for the electric dipole's charge separation k, and the unit cell volume V_c. This combination of physical parameters relates a modulated electric dipole to the polarization-adjusted electric potential that it produces.

A material's dc polarization is proportional to $1 - 1/\varepsilon_0$. This polarization contains contributions from displacing electrons as well as from displacing relatively massive and slow-moving ions. By comparison the optical polarization $1 - 1/\varepsilon_\infty$ only contains contributions from electrons since only they can move fast enough to follow a high-frequency oscillating electric field. Thus, $1/\varepsilon_\infty - 1/\varepsilon_0$ measures the slow polarization arising from displacing ions. This difference of the reciprocal of the dielectric constants tends to be small for covalent materials since there $\varepsilon_0 \approx \varepsilon_\infty$. Hence the associated long-range electron–phonon coupling also tends to be small for covalent materials. By contrast, the long-range electron–phonon coupling is of major importance in ionic crystals such as alkali halides when $\varepsilon_0 \approx 2\varepsilon_\infty$. The long-range electron–phonon coupling is exceptionally strong in ferroelectric and extremely polar materials as they are characterized by $\varepsilon_0 \gg \varepsilon_\infty$.

The potential well produced by ions' relaxation about a static point charge is a useful fiducial situation when discussing self-trapping. Ions assuming shifted equilibrium positions in response to the presence of the static point charge produce a Coulomb-like well

$$V_{eq}^{LR}(\boldsymbol{r}) = -\left(\frac{1}{\varepsilon_\infty} - \frac{1}{\varepsilon_0} \right) \frac{e^2}{|\boldsymbol{r}|}. \tag{2.3}$$

The magnitude of this Coulombic potential measures the strength of the long-range electron–phonon interaction. The potential well generated by the long-range electron–phonon interaction about a confined carrier of radius R is schematically illustrated in Fig. 2.2. As discussed in Section 4.3, the well depth is proportional to $1/R$.

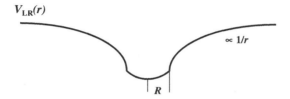

Fig. 2.2 A potential well forms about a confined carrier of radius R as its long-range electron–phonon interaction induces shifts of the equilibrium positions of surrounding ions.

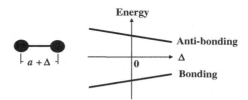

Fig. 2.3 The energies of the bonding and anti-bonding states of a covalent bond change as its length shifts from its carrier-free equilibrium value a by Δ.

2.2 Short-range electron–phonon interactions

The short-range electron–phonon interaction results when the energy of a carrier depends on the strain where it resides. Within the continuum treatment the short-range electron–phonon interaction is represented by the contact interaction

$$Z_{SR}(\boldsymbol{r}-\boldsymbol{u}) \equiv -F\delta(\boldsymbol{r}-\boldsymbol{u}), \tag{2.4}$$

where the constant F has the dimensions of a force. Strain-induced shifts of electronic transfer energies and the energies of bonding and anti-bonding states contribute to the short-range electron–phonon interaction. For example, as indicated in Fig. 2.3, compressing a covalent bond tends to lower the energy of the bonding state while increasing the energy of the anti-bonding state.

The lowering of the effective potential for a carrier on a single isolated structural unit (e.g. a bond) is used to reference the strength of the short-range interaction. Within the continuum treatment the carrier-induced strain is that which minimizes the sum of the carrier's energy, $-F\Delta(\boldsymbol{r})$, plus the Hooke's-law strain energy for the associated strain, $k\Delta(\boldsymbol{r})^2/2$, where k denotes the continuum's stiffness constant. The equilibrium strain is thus $\Delta_{eq} = F/k$. The lowering of the potential energy of Eq. (2.1) when a carrier is confined to a single bond is

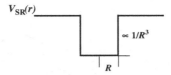

Fig. 2.4 A potential well forms about a confined carrier of radius R as its short-range electron–phonon interaction induces shifts of the equilibrium positions of atoms it contacts.

$$V_{eq}^{SR} = -\frac{F^2}{k}. \tag{2.5}$$

The potential well generated by the short-range electron–phonon interaction about a confined carrier of radius R is schematically illustrated in Fig. 2.4. As discussed in Section 4.3, the well depth is proportional to $1/R^3$.

The depth of this potential is typically less than 1 eV since values of F are usually 2–3 eV/Å with $k > 10$ eV/Å^2 for a covalent bond. By comparison, the depth of the potential produced by the long-range interaction in an ionic material, Eq. (2.3), is about 3 eV with $\varepsilon_0 \approx 2\varepsilon_\infty = 5$ and $r \approx 1$ Å. Since ionic and polar materials possess relatively strong electron–phonon interactions their effects have become known as *polaron* effects. Nonetheless, polaron effects arising from short-range electron–phonon interactions can be significant in ionic materials as well as in covalent solids.

3

Weak-coupling polarons: carrier-induced softening

With a sufficiently weak electron–phonon interaction a charge carrier in a semi-conductor forms a weak-coupling polaron. The energies and the effective masses associated with states located near the extrema of a carrier's energy band are thereby shifted modestly. These effects can be calculated by treating the electron–phonon interaction as a small perturbation that links a carrier in a wide electronic energy band, $W \gg \hbar\,\omega$, with the harmonic vibrations of the solid's atoms. This perturbation scheme is valid provided that the electron–phonon interaction is too weak and the electronic energy band too wide for the carrier to self-trap.

3.1 Long-range electron–phonon interaction

With the long-range electron–phonon interaction of ionic and polar materials the carrier interacts primarily with long-wavelength phonons. In the absence of thermal phonons the energy of an electron at the bottom of a conduction band is then shifted by

$$\Delta E_{\text{WC}}^{\text{LR}} = -\frac{e^2}{2R_{\text{p}}} \left(\frac{1}{\varepsilon_\infty} - \frac{1}{\varepsilon_0} \right). \tag{3.1}$$

Here the "weak-coupling polaron radius" R_{p} is defined in terms of the carrier's bare mass m, the lattice constant a, and the associated electronic bandwidth parameter $t \equiv \hbar^2/2ma^2$ by

$$R_{\text{p}} \equiv \left(\frac{\hbar}{2m\omega} \right)^{1/2} = \left(\frac{t}{\hbar\omega} \right)^{1/2} a, \tag{3.2}$$

and $W = 12t$ is the width of a nearest-neighbor tight-binding electronic band of a cubic crystal. This so-called polaron radius is not the radius of a self-trapped state. In fact, the weak-coupling polaron radius is independent of the electron–phonon coupling. This radius just describes a distance that a carrier can diffuse within the period of an atomic vibration.

This weak-coupling approach breaks down when the electron–phonon interaction strength and the electronic bandwidth are such as to produce a self-trapping potential well that can bind a carrier with an energy that exceeds the characteristic (optical) phonon energy (Fröhlich, 1963). The carrier can then circulate within its self-trapping potential well with a frequency, the binding energy divided by Planck's constant, that exceeds the phonon frequency.

The long-range electron–phonon coupling is often expressed in terms of the dimensionless Fröhlich coupling constant (Fröhlich, 1963):

$$\alpha \equiv \frac{e^2}{\hbar}\left(\frac{1}{\varepsilon_\infty}-\frac{1}{\varepsilon_0}\right)\sqrt{\frac{m}{2\hbar\omega}}. \tag{3.3}$$

As will become clear in Chapter 4, the Fröhlich coupling constant is essentially the square root of the ratio of the strong-coupling binding energy, that with the carrier being self-trapped, to the phonon energy. Thus, the weak-coupling regime, characterized by $\alpha < 1$, indicates that the long-range electron–phonon coupling is too weak to produce self-trapping. Expressed in terms of α, the zero-temperature weak-coupling polaron energy associated with the long-range electron–phonon coupling given by Eq. (3.1) is simply

$$\Delta E_{\mathrm{WC}}^{\mathrm{LR}} = -\alpha\hbar\omega. \tag{3.4}$$

The corresponding effective mass for a weak-coupling polaron is also readily described in terms of α:

$$\frac{m^*}{m} = 1 + \frac{\alpha}{6}. \tag{3.5}$$

Since $\alpha < 1$ it is evident that the effective mass of a weak-coupling polaron is only slightly higher than that of the bare carrier.

3.2 Short-range electron–phonon interactions

The short-range electron–phonon interaction also lowers the energy of a carrier at the bottom of a conduction band. However, the short range of the electron–phonon interaction fosters a more effective interaction with short-wavelength phonons than occurs with the long-range electron–phonon interaction. Treating the short-range electron–phonon interaction as a perturbation gives the corresponding zero-temperature weak-coupling polaron energy:

$$\Delta E_{\text{WC}}^{\text{SR}} = -\frac{F^2/k}{12\left(R_{\text{p}}/a\right)^2}.$$ (3.6)

The difference in the ranges of their electron–phonon interactions accounts for the different dependences of $\Delta E_{\text{WC}}^{\text{LR}}$ and $\Delta E_{\text{WC}}^{\text{SR}}$ on R_{p} that are manifested between Eq. (3.1) and Eq. (3.6).

The condition governing breakdown of a short-range electron–phonon interaction's weak-coupling solution yields a condition that can be understood with a classical argument (Emin, 1973a). In particular, the weak-coupling solution fails if the product of the force that a carrier exerts on adjacent atoms multiplied by the time that a carrier can spend at a site exceeds the momentum of these atoms' vibrations: $F(\hbar/t) > P_{\text{vib}}$. Atoms' vibratory momentum increases with rising temperature, thereby increasing the stability of the weak-coupling solution. At zero temperature, with only zero-point vibrations, the weak-coupling solution fails when $[(F^2/k)\hbar\omega]^{1/2} > t$.

3.3 Two-site model: origin of carrier-induced softening

A general feature of the weak-coupling polaron is that its energy vanishes in the limit that atoms' vibration frequencies vanish. In particular, $\Delta E_{\text{WC}}^{\text{LR}}$ and $\Delta E_{\text{WC}}^{\text{SR}}$ of Eq. (3.1) and Eq. (3.6) vanish since R_{p} tends to infinity with diminishing atomic-vibration frequency. By contrast, the energy lowering associated with self-trapping, discussed in Chapter 4, will be seen to remain finite in this limit. This distinction indicates a fundamental difference in the physics underlying weak-coupling and strong-coupling polarons.

To elucidate this difference consider the model of Fig. 3.1 in which an electron transfers between two independent equivalent harmonically vibrating molecules. Deformation of each of the two molecules is characterized by a single distortion parameter, Δ_i with $i = 1$ or 2. The Hamiltonian describing carrier-free harmonic vibrations of these molecules is:

$$H_{\text{vib}} = -\frac{\hbar^2}{2M}\left(\frac{\partial^2}{\partial\Delta_1^2} + \frac{\partial^2}{\partial\Delta_2^2}\right) + \frac{k}{2}\left(\Delta_1^2 + \Delta_2^2\right),$$ (3.7)

where M is the reduced mass associated with a molecule's vibrations and k is the associated spring constant. A short-range electron–phonon interaction results from

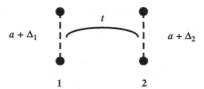

Fig. 3.1 An electron transfers with energy t between two molecules whose deformation parameters are Δ_1 and Δ_2, respectively.

the energy of the electron occupying a molecule depending upon the deformation of that molecule. With the electron occupying the i-th molecule:

$$\varepsilon_{\mathrm{SR},i} = -F\Delta_i. \tag{3.8}$$

The electron's intermolecular motion is associated with the transfer energy t.

In the spirit of the tight-binding approach, an electronic eigenstate is expressed as a superposition of local states associated with its occupation of each of the two molecules:

$$\Psi = a_1\Phi_1 + a_2\Phi_2. \tag{3.9}$$

The expansion coefficients are found by solving the eigenvalue matrix equation:

$$\begin{bmatrix} -F\Delta_1 - W & t \\ t & -F\Delta_2 - W \end{bmatrix} \begin{bmatrix} a_1 \\ a_2 \end{bmatrix} = 0. \tag{3.10}$$

The roots of the associated determinant are the two electronic eigenvalues:

$$W_{\pm} = -\frac{F(\Delta_2 + \Delta_1)}{2} \pm \sqrt{\left[\frac{F(\Delta_2 - \Delta_1)}{2}\right]^2 + t^2}. \tag{3.11}$$

In the absence of intermolecular transfer, $t = 0$, the eigenvalues are just the electronic energies for the electron confined on one or the other of the two molecules: $W = -F\Delta_1$ and $W = -F\Delta_2$. In the complementary limit of vanishing electron–phonon interaction, $F = 0$, the eigenvalues are t and $-t$, the upper and lower states of the two-state energy band associated with intermolecular transfer.

The molecule occupied by the electron is governed by the relative values of their deformation parameters. It is therefore expeditious to introduce the relative deformation parameter $x \equiv (\Delta_2 - \Delta_1)/\sqrt{2}$ along with the complementary parameter

$X \equiv (\Delta_2 + \Delta_1)/\sqrt{2}$. Rewriting the vibration Hamiltonian of Eq. (3.7) and the electronic eigenvalues of Eq. (3.11) in terms of these parameters yields:

$$H_{vib} = -\frac{\hbar^2}{2M}\left(\frac{\partial^2}{\partial X^2} + \frac{\partial^2}{\partial x}\right) + \frac{k}{2}\left(X^2 + x^2\right) \tag{3.12}$$

and

$$W_{\pm} = -\frac{FX}{\sqrt{2}} \pm \sqrt{\frac{(Fx)^2}{2} + t^2}. \tag{3.13}$$

The x-dependent portion of the sum of the vibration and electronic potentials governs the electron's intermolecular motion:

$$V_{\pm}(x) \equiv \frac{k}{2}x^2 \pm \sqrt{\frac{(Fx)^2}{2} + t^2}. \tag{3.14}$$

As illustrated in Fig. 3.2, increasing the strength of the electron–phonon interaction softens the potential of the lower electronic state $V_-(x)$ and stiffens the potential of the upper electronic state $V_+(x)$ (Öpik and Pryce, 1957).

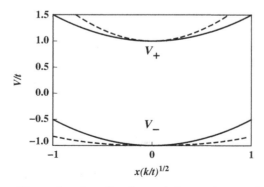

Fig. 3.2 The sum of the molecular-vibration and electronic energies (in units of the electronic transfer energy t) is plotted against the relative deformation parameter x (in units of $\sqrt{t/k}$) for the two electronic branches. Increasing the electron–phonon coupling strength from zero (solid curves) to a finite value (dashed curves) reduces the curvature of the lower electronic branch, V_-, and increases the curvature of the upper electronic branch, V_+.

Both potentials remain harmonic for sufficiently weak electron–phonon coupling:

$$V_\pm(x) \cong \pm t + \frac{[k \pm (F^2/2t)]}{2}x^2.$$

(3.15)

Softening of the harmonic ground-state potential reduces its zero-point energy by

$$\Delta E = \frac{\hbar}{2}\left(\sqrt{\frac{k-F^2/2t}{M}} - \sqrt{\frac{k}{M}}\right)$$

(3.16)

$$= \frac{\hbar\omega}{2}\left(\sqrt{1-F^2/2kt} - 1\right) \cong -\frac{F^2/k}{8(R_p/a)^2},$$

where R_p is defined in Eq. (3.2) and the numerical coefficient for this two-site model differs by a factor of 3/2 from that of Eq. (3.6) obtained for a three-dimensional cubic lattice. In a similar manner, stiffening of the potential for the upper state increases its zero-point energy by $|\Delta E|$. The energy separation between the lower and upper state is thereby increased. More generally, as depicted in Fig. 3.3, the electron–phonon interaction produces a temperature increase of the width of an energy band comprising weak-coupling states.

The processes that produce this softening and stiffening can be understood by considering how the electron moves in response to atomic vibrations. Solution of Eq. (3.10) describes the transfer of the electron upon changing the relative molecular deformation parameter:

$$P_\pm(y) \equiv |a_2^\pm|^2 - |a_1^\pm|^2 = \frac{\left(-y \pm \sqrt{1+y^2}\right)^2 - 1}{\left(-y \pm \sqrt{1+y^2}\right)^2 + 1} \cong \mp y,$$

(3.17)

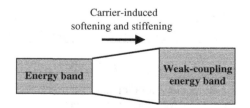

Carrier-induced
softening and stiffening

Fig. 3.3 Electron–phonon coupling lowers the interacting system's energy for low-energy electronic states and raises the interacting system's energy for high-energy electronic states. The electron–phonon interaction thereby broadens the band of system energies.

where $y \equiv Fx/\sqrt{2}t = F(\Delta_2-\Delta_1)/2t = (\varepsilon_{SR,1} - \varepsilon_{SR,2})/2t$. As illustrated in Fig. 3.4 for the (negative) root associated with softening, the electron moves from molecule 1 to molecule 2 as its electronic energy falls below that of molecule 1. By contrast, for the (positive) root associated with stiffening, the electron moves from molecule 2 to molecule 1 despite its electronic energy rising above that of molecule 2. That is, softening occurs as the electron moves toward the lower energy site thereby enhancing its molecule's deformation. However, stiffening occurs when the electron moves away from the lower energy site thereby reducing its molecule's deformation.

The harmonic approximation of Eq. (3.15) is valid if the amplitude of a zero-point vibration is less than the maximum value of x for which the potential of Eq. (3.14) remains quadratic: $\sqrt{\hbar\omega/k} < \sqrt{2}t/F$. This condition, $\sqrt{(F^2/2k)\hbar\omega} < t$, is essentially just that for the validity of the zero-temperature weak-coupling polaron in a cubic crystal (Emin, 1972; Emin 1973a). In essence, the weak-coupling solution describes a carrier whose adjustment to displacements of atoms just modifies the stiffness constants that govern their vibrations.

A different type of solution, corresponding to a self-trapped carrier, emerges when $F^2/2k > t$. In this circumstance, as shown in Fig. 3.5, $V_-(x)$ of Eq. (3.14) garners two minima (Öpik and Pryce, 1957). As $F^2/2kt$ is increased the two minima move from $x \approx 0$ toward $x = \pm F/\sqrt{2}k$. Deforming the molecules to establish one of these minima corresponds to localizing the electron on one of the two molecules while lowering its ground-state energy by $F^2/2k$.

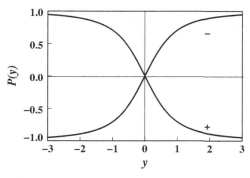

Fig. 3.4 The probability of the electron occupying molecule 2 versus occupying molecule 1 is plotted for positive and negative branches of the electronic spectrum. $P(y) > 0$ indicates that molecule 2 is preferentially occupied and $P(y) < 0$ indicates that molecule 1 is preferentially occupied. Occupation of molecule 1 or 2 is energetically preferred for $y < 0$ and $y > 0$, respectively. Thus the negative branch has the electron occupying the energetically preferred molecule. However, the positive branch has the carrier occupying the energetically unfavorable molecule.

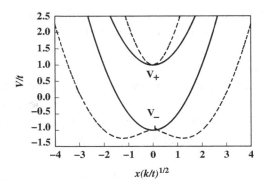

Fig. 3.5 The sum of the molecular-vibration and electronic energies (in units of the electronic transfer energy t) is plotted against the relative deformation parameter x (in units of $\sqrt{t/k}$) for the two electronic branches. Increasing the electron–phonon coupling strength from zero (solid curves) to a large enough value (dashed curves) increases the curvature of the upper electronic branch, V_+, while producing two symmetrically placed self-trapping minima on the lower electronic branch, V_-.

This section illustrated carrier-induced softening by considering a charge carrier that can move between two sites. However, this example does not correctly depict self-trapping. Self-trapping in a solid cannot be adequately modeled with just two sites. The following chapters will address self-trapping in considerably more detail.

The adiabatic limit reveals a qualitative difference between a weak-coupling polaron and a strong-coupling polaron. The adiabatic limit envisions letting atomic masses approach infinity while keeping interatomic stiffness constants finite. Atomic vibration frequencies vanish in the adiabatic limit. The energy lowering of a weak-coupling polaron results because a carrier's interactions with atoms alter their vibration frequencies. Therefore a weak-coupling polaron's energy lowering vanishes in the adiabatic limit. By contrast, a strong-coupling polaron is based on the self-trapping that occurs as a carrier's interactions with atoms shift their equilibrium positions. These shifts depend on the stiffness constants governing the associated atomic displacements. The energy lowering of a strong-coupling polaron remains finite in the adiabatic limit.

4

Strong coupling: self-trapping

This chapter presents the theory of an electron's self-trapping. The discussion begins by reviewing the adiabatic formalism. This formalism provides a method of describing an electron amidst atoms with which it interacts. In Section 4.2 the formalism is applied to the nonlinear problem of an electron's self-localization. A scaling approach is then employed in Section 4.3 to analyze an electron's self-trapping in the adiabatic limit in which atoms' masses are assumed to be infinite. Two distinct types of self-trapped state, corresponding to large- and small-polaron formation, are possible. The situations in which large and small polarons can form are described. The discussion proceeds beyond the adiabatic limit in Section 4.4 to describe the effects of atomic vibrations on self-trapped carriers and on the conditions of their formation. The chapter concludes by describing how disorder can induce electrons to collapse into small polarons.

4.1 Adiabatic formalism

An electron is very much lighter than an atom. This basic feature generates a qualitative asymmetry in the responses of an electronic carrier and a solid's atoms to their mutual interactions. The adiabatic approach exploits this asymmetry to describe the motion of an electronic carrier in a solid (Born and Oppenheimer, 1927).

The Hamiltonian for the added electron and the solid is written as

$$H \equiv H_e(\boldsymbol{r}, \boldsymbol{\Delta}) + H_{vib}(\boldsymbol{\Delta}). \tag{4.1}$$

The electronic portion of this Hamiltonian, $H_e(\boldsymbol{r}, \boldsymbol{\Delta})$, describes the carrier's kinetic energy and the potential arising from the carrier's interaction with the solid's atoms. The carrier's position is designated by $\boldsymbol{r}$, and $\boldsymbol{\Delta}(\boldsymbol{u})$ represents the displacement of the atom whose equilibrium position in a carrier-free solid is $\boldsymbol{u}$. Here, for notational simplicity, the set of atomic displacements is collectively represented by the single atomic-deformation function $\Delta \equiv \Delta(\boldsymbol{u})$. The second contribution to the total Hamiltonian, $H_{vib}(\Delta)$, describes the kinetic energy of the atoms and their interactions with one another. In polaron studies this Hamiltonian usually describes harmonic vibrations of a solid's atoms.

The adiabatic approach is based on the eigenstates of an electron moving amongst atoms whose positions are fixed:

$$H_e(\boldsymbol{r}, \Delta)\phi_n(\boldsymbol{r}, \Delta) = W_n(\Delta)\phi_n(\boldsymbol{r}, \Delta). \tag{4.2}$$

Here $\phi_n(\boldsymbol{r},\Delta)$ and $W_n(\Delta)$ represent the n-th electronic eigenstate and corresponding energy for an electron in a medium whose atomic-displacement vectors are collectively designated as Δ. However, the excess electron also affects atomic motion. In particular, atomic motion is described by the projection of the total Hamiltonian on the electronic state $\phi_n(\boldsymbol{r},\Delta)$:

$$\left[\int d\boldsymbol{r}\phi_n^*(\boldsymbol{r}, \Delta)H\phi_n(\boldsymbol{r}, \Delta)\right]\chi_m^n(\Delta) = E_{n,m}\chi_m^n(\Delta), \tag{4.3}$$

where $\chi_m^n(\Delta)$ and $E_{n,m}$ depict the m-th vibration wavefunction and the corresponding total energy when the electron is in its n-th electronic state.

The left-hand side of Eq. (4.3) can be evaluated further with

$$H_{vib}(\Delta) = \frac{\hat{\boldsymbol{P}}^2}{2M} + V_{vib}(\Delta), \tag{4.4}$$

where $\hat{\boldsymbol{P}} \equiv (\hbar/i)\partial/\partial\Delta$ represents the atomic momentum operator and M is the appropriate reduced atomic mass. After some manipulations, the Hamiltonian governing atoms' motion can be expressed as (Emin and Holstein, 1969):

$$H_{ad} = \frac{\left[\hat{\boldsymbol{P}}^2 + q\left(\boldsymbol{A}\cdot\hat{\boldsymbol{P}} + \hat{\boldsymbol{P}}\cdot\boldsymbol{A}\right) - \hbar^2 \operatorname{Re}\int d\boldsymbol{r}\phi_n^*(\boldsymbol{r}, \Delta)\dfrac{\partial^2\phi_n(\boldsymbol{r}, \Delta)}{\partial\Delta^2}\right]}{2M}$$
$$+ V_{vib}(\Delta) + W_n(\Delta), \tag{4.5}$$

where a portion of the Hamiltonian's inertial contribution has been written in terms of a fictitious vector potential for an electronic carrier of charge q:

$$qA(\Delta) \equiv \frac{\hbar}{i}\int d\boldsymbol{r}\phi_n^*(\boldsymbol{r}, \Delta)\frac{\partial\phi_n(\boldsymbol{r}, \Delta)}{\partial\Delta}. \tag{4.6}$$

The final term in the square brackets is usually dismissed as being relatively small.

In the absence of a magnetic field a non-degenerate electronic state can be taken to be real. The fictitious vector potential then vanishes since

$$\int dr \phi_n(r, \Delta) \frac{\partial \phi_n(r, \Delta)}{\partial \Delta} = \frac{1}{2} \frac{\partial}{\partial \Delta} \int dr \phi_n^2(r, \Delta) = 0. \tag{4.7}$$

With a magnetic field $\phi_n(r, \Delta)$ becomes complex and $A(\Delta)$ no longer vanishes. The effective magnetic field obtained from $A(\Delta)$ is proportional to the magnetic field applied to the electron. The adiabatic Hamiltonian of Eq. (4.5) then describes how the circulation of the electron in a magnetic field affects atoms' vibrations. Such considerations are central to the adiabatic theory of the polaron Hall effect (Emin and Holstein, 1969).

An electron can move adiabatically in response to atomic motion in accord with $\phi_n(r, \Delta)$. However, atomic motion can also induce transitions between adiabatic states: $\phi_n(r, \Delta) \rightarrow \phi_{n'}(r, \Delta)$. The matrix element that governs such non-adiabatic transitions is

$$\frac{\hat{P}}{M} \cdot \frac{\hbar}{i} \int dr \phi_{n'}^*(r, \Delta) \frac{\partial \phi_n(r, \Delta)}{\partial \Delta} - \frac{\hbar^2}{2M} \int dr \phi_{n'}^*(r, \Delta) \frac{\partial^2 \phi_n(r, \Delta)}{\partial \Delta^2}. \tag{4.8}$$

The first of these two terms typically dominates. Its magnitude is the product of an atomic velocity and $\hbar$ divided by the distance atoms must move to appreciably alter the electronic state ($\sim t/F$ with a short-range electron–phonon interaction). Non-adiabatic transitions only become important when the product of the relevant atomic velocity and the electronic matrix element exceeds $W_{n'}(\Delta) - W_n(\Delta)$. In most semiconducting materials non-adiabatic transitions can be ignored since atoms' vibration energies are usually much smaller than the relevant electronic energies. Thus the adiabatic approach is frequently adopted.

4.2 Self-trapping

The adiabatic approach is widely utilized to address self-trapping. Here the adiabatic method is used to study the self-trapping of an electron added to a deformable continuum. A linear electron–phonon interaction is introduced by taking the electron's potential energy to be proportional to the medium's deformations. These deformations, like a solid's dilation, are taken as scalar parameters attributed to positions within the continuum that are denoted by the vector u. The continuum's strain energy is a quadratic function of these deformations.

The adiabatic potential for the electron and its deformable continuum is the sum of the electron's energy and the strain energy of the continuum. Finding the minimum of the adiabatic energy involves calculating the first partial derivative of the electron's energy with respect to the deformation parameters, the $\Delta(u)$. This differentiation is equivalent to calculating the expectation value of the derivative of the Hamiltonian with respect to that deformation:

$$\frac{\partial W_n}{\partial \Delta} = \frac{\partial}{\partial \Delta} \langle \phi_n(r, \Delta)|H_e|\phi_n(r, \Delta)\rangle$$

$$= W_n \frac{\partial}{\partial \Delta} \langle \phi_n(r, \Delta)|\phi_n(r, \Delta)\rangle + \langle \phi_n(r, \Delta)\left|\frac{\partial H_e}{\partial \Delta}\right|\phi_n(r, \Delta)\rangle$$

$$= \langle \phi_n(r, \Delta)\left|\frac{\partial H_e}{\partial \Delta}\right|\phi_n(r, \Delta)\rangle, \tag{4.9}$$

where the normalization condition $\langle \phi_n(r, \Delta)|\phi_n(r, \Delta)\rangle = 1$ has been utilized.

The Hamiltonian for the electron in the deformable continuum is

$$H_e = \frac{-\hbar^2}{2m} \nabla_r^2 + \int du Z(r - u)\Delta(u), \tag{4.10}$$

where m is the electron's mass and $Z(r - u)$ describes the magnitude and range of the electron–phonon interaction. The partial derivative of the electronic energy with respect to $\Delta(u)$ is therefore

$$\frac{\partial W_n}{\partial \Delta(u)} = \int dr |\phi_n(r, \Delta)|^2 Z(r - u). \tag{4.11}$$

The strain energy of the carrier-free harmonic deformable continuum is

$$V_{\text{strain}} = \frac{S}{2} \int du \Delta^2(u), \tag{4.12}$$

where S is the continuum's stiffness per unit volume. Equations (4.11) and (4.12) are used to find the deformation pattern for which the first derivative of the adiabatic potential vanishes:

$$\Delta_{\text{eq}}(u) = \frac{-\int dr |\phi_n(r, \Delta_{\text{eq}})|^2 Z(r - u)}{S}. \tag{4.13}$$

This relation provides an implicit equation for the equilibrium deformation surrounding the electron. The equilibrium deformation surrounding the carrier depends on (1) the continuum's stiffness, (2) the magnitude and range of the electron–phonon interaction, and (3) the self-trapped electron's spatial extent.

The electron also alters the stiffness associated with deforming the medium surrounding it. Elements of the stiffness tensor linking deformations $\Delta(\boldsymbol{u})$ and $\Delta(\boldsymbol{u}')$ in the presence of the added electron, $S_e(\boldsymbol{u},\boldsymbol{u}')$, are given by second partial derivatives of the adiabatic potential with respect to these deformation parameters:

$$S_e(\boldsymbol{u}, \boldsymbol{u}') = S\delta(\boldsymbol{u} - \boldsymbol{u}') + \int d\boldsymbol{r} \frac{\partial |\phi_n(\boldsymbol{r}, \Delta)|^2}{\partial \Delta(\boldsymbol{u}')} Z(\boldsymbol{r} - \boldsymbol{u}). \qquad (4.14)$$

Equation (4.14) shows that carrier-induced changes of the system's stiffness result from the wavefunction of an electron changing as the continuum is deformed. The wavefunction changes because deforming the continuum alters the potential well experienced by the electron. Figure 4.1 illustrates the adjustment of a localized electron in response to displacing a solid's ions from their minimum-energy configuration. The electron adjusts as the ions shift so as to minimize its energy. Since this electronic relaxation lowers the net energy required to produce the deformation, the associated stiffness is reduced. This phenomenon is termed *carrier-induced softening* (Emin, 1994a; Emin, 1994b; Emin, 2000a).

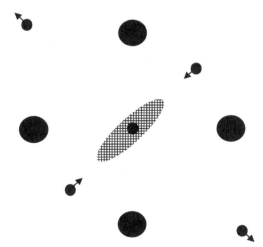

Fig. 4.1 Asymmetrical displacements of cations from their equilibrium positions (small dots with arrows) asymmetrically deform the charge distribution of the self-trapped electron (hatched region) that occupies the central cation. The large dots are anions that remain fixed at their equilibrium positions.

This carrier-induced softening may be explicitly expressed in terms of the electron–phonon interaction by using second-order perturbation theory to describe the deformation-induced change of the electronic wavefunction:

$$S_{\mathrm{e}}(\boldsymbol{u},\boldsymbol{u}') = S\delta(\boldsymbol{u}-\boldsymbol{u}')$$

$$-\sum_{n'\neq n}\frac{\int d\boldsymbol{r}'\phi_n^*(\boldsymbol{r}')Z(\boldsymbol{r}'-\boldsymbol{u}')\psi_{n'}(\boldsymbol{r}')\int d\boldsymbol{r}\psi_{n'}^*(\boldsymbol{r})Z(\boldsymbol{r}-\boldsymbol{u})\phi_n(\boldsymbol{r}) + \text{c.c.}}{E_{n'}-W_n}. \qquad (4.15)$$

Here W_n and ϕ_n designate the energy and electronic wavefunction for the ground state while $E_{n'}$ and $\psi_{n'}$ refer to the energies and corresponding wavefunctions of excited electronic states. Carrier-induced softening is evident since $E_{n'} > W_n$. Equation (4.15) explicitly manifests the symmetry of the stiffness tensor with respect to interchange of $\boldsymbol{u}$ and $\boldsymbol{u}'$.

Carrier-induced stiffness changes can also be related to the self-trapped electron's polarizability tensor. Inserting expansions of the electron–phonon interaction function about the electron's centroid, $\boldsymbol{r}_{\mathrm{c}}$:

$$Z(\boldsymbol{r}-\boldsymbol{u}) \cong Z(\boldsymbol{r}_{\mathrm{c}}-\boldsymbol{u}) + \frac{\partial Z(\boldsymbol{r}_{\mathrm{c}}-\boldsymbol{u})}{\partial \boldsymbol{r}_{\mathrm{c}}}\cdot(\boldsymbol{r}-\boldsymbol{r}_{\mathrm{c}})$$

$$= Z(\boldsymbol{r}_{\mathrm{c}}-\boldsymbol{u}) - \frac{\partial Z(\boldsymbol{r}_{\mathrm{c}}-\boldsymbol{u})}{\partial \boldsymbol{u}}\cdot(\boldsymbol{r}-\boldsymbol{r}_{\mathrm{c}}) \qquad (4.16)$$

into Eq. (4.15) enables it to be rewritten as

$$S_{\mathrm{e}}(\boldsymbol{u},\boldsymbol{u}') = S\delta(\boldsymbol{u}-\boldsymbol{u}') - \frac{\partial Z(\boldsymbol{r}_{\mathrm{c}}-\boldsymbol{u}')}{\partial \boldsymbol{u}'}\cdot\overset{\leftrightarrow}{P}\cdot\frac{\partial Z(\boldsymbol{r}_{\mathrm{c}}-\boldsymbol{u})}{\partial \boldsymbol{u}}, \qquad (4.17)$$

where the electronic polarizability tensor is

$$\overset{\leftrightarrow}{P} \equiv \sum_{n'\neq n}\frac{\int d\boldsymbol{r}'\phi_n^*(\boldsymbol{r}')(\boldsymbol{r}'-\boldsymbol{r}_{\mathrm{c}})\psi_{n'}(\boldsymbol{r}')\int d\boldsymbol{r}\phi_{n'}^*(\boldsymbol{r})(\boldsymbol{r}-\boldsymbol{r}_{\mathrm{c}})\psi_n(\boldsymbol{r}) + \text{c.c.}}{E_{n'}-W_n}. \qquad (4.18)$$

The electronic polarizability tends to increase strongly with the size of the electronic state. A simple model gives the polarizability increasing in proportion to the fourth power of the state's radius. As a result, carrier-induced softening may be especially important in molecular solids when an electronic state is comparable in size to a molecule.

Nonetheless, a carrier confined to a single molecule of Holstein's pedagogic molecular-crystal model does not soften its vibrations (Holstein, 1959a). Rather the model implicitly represents the localized carrier as a non-polarizable fixed point. Indeed, the model's pseudo-molecule is designed to depict the situation envisioned in simple transition-metal oxides: large displaceable anions surrounding a small fixed cation that can be occupied by an electronic hole.

The ground-state values of the electronic energy and wavefunction, W_n and $\phi_n(r)$, are obtained when the continuum deforms to its carrier-induced equilibrium configuration. In particular, the potential in the Hamiltonian for the ground-state eigenvalue equation is that produced by the equilibrium deformation of Eq. (4.13). Since the equilibrium deformation depends on the electron's wavefunction, the resulting eigenvalue equation is nonlinear:

$$\left[\frac{-\hbar^2}{2m}\nabla_r^2 - \frac{1}{S}\int dr'|\phi_n(r')|^2\int du Z(r-u)Z(r'-u)\right]\phi_n(r) = W_n\phi_n(r). \qquad (4.19)$$

Solutions of this differential equation depend qualitatively on the range of the electron–phonon interaction and on the electron's dimensionality. That is, the solution depends on whether the electron can move in all three dimensions or is restricted to a plane or to a chain. These features are discussed in Section 4.3.

The energy of a polaron can be written as the sum of three contributions: (1) the ground-state's electronic energy, (2) the associated deformational energy, and (3) the change of vibrations' free energy resulting from carrier-induced alterations of their frequencies. The ground-state's electronic energy is found from Eq. (4.19) to be:

$$W_n = \frac{\hbar^2}{2m}\int dr|\nabla_r\phi_n(r)|^2 - \frac{1}{S}\int dr|\phi_n(r)|^2\int dr'|\phi_n(r')|^2\int du Z(r-u)Z(r'-u). \qquad (4.20)$$

The strain energy needed to produce the ground-states' equilibrium deformation is

$$\frac{S}{2}\int du\Delta_{eq}^2(u) = \frac{1}{2S}\int dr|\varphi_n(r)|^2\int dr'|\varphi_n(r')|^2\int du Z(r-u)Z(r'-u). \qquad (4.21)$$

This deformation energy has the same form as the deformation-induced lowering of the electron's potential energy. Thus these two contributions are combined in writing the minimum of the adiabatic potential of Eq. (4.5) with $\Delta \equiv \Delta_{eq}$ as:

$$V_{ad}^{min} = \frac{\hbar^2}{2m}\int dr|\nabla_r\phi_n(r)|^2 - \int dr|\phi_n(r)|^2\int dr'|\phi_n(r')|^2 I(r,r'), \qquad (4.22)$$

where

$$I(\boldsymbol{r}, \boldsymbol{r}') \equiv \frac{1}{2S} \int d\boldsymbol{u} Z(\boldsymbol{r} - \boldsymbol{u}) Z(\boldsymbol{r}' - \boldsymbol{u}). \tag{4.23}$$

The interaction function $I(\boldsymbol{r},\boldsymbol{r}')$ has been evaluated for models of the electron–phonon interaction. For the Fröhlich long-range electron–phonon interaction the interaction function is

$$I_{\mathrm{LR}}(\boldsymbol{r}, \boldsymbol{r}') = \left(\frac{1}{\varepsilon_\infty} - \frac{1}{\varepsilon_0} \right) \frac{e^2}{2|\boldsymbol{r} - \boldsymbol{r}'|}. \tag{4.24}$$

The interaction function for the short-range electron–phonon interaction of Eq. (2.4) is readily calculated to yield

$$I_{\mathrm{SR}}(\boldsymbol{r}, \boldsymbol{r}') = \frac{F^2}{2k} V_{\mathrm{c}} \delta(\boldsymbol{r} - \boldsymbol{r}') \tag{4.25}$$

where $S \equiv k/V_{\mathrm{c}}$ and V_{c} is the unit cell volume. Upon summing these long-range and short-range components of the interaction function, the adiabatic potential, Eq. (4.22), becomes

$$V_{\mathrm{ad}}^{\mathrm{min}} = \frac{\hbar^2}{2m} \int d\boldsymbol{r} |\nabla_{\boldsymbol{r}} \phi_n(\boldsymbol{r})|^2$$

$$- \frac{e^2}{2} \left(\frac{1}{\varepsilon_\infty} - \frac{1}{\varepsilon_0} \right) \int d\boldsymbol{r} \int d\boldsymbol{r}' \frac{|\phi_n(\boldsymbol{r})|^2 |\phi_n(\boldsymbol{r}')|^2}{|\boldsymbol{r} - \boldsymbol{r}'|} - \left(\frac{F^2}{2k} \right) V_{\mathrm{c}} \int d\boldsymbol{r} |\phi_n(\boldsymbol{r})|^4. \tag{4.26}$$

The third contribution to a polaron's energy is produced by its altering of the frequencies with which the atoms surrounding it oscillate. The associated change of the free energy of these atoms' vibrations is

$$\Delta F_{\mathrm{vib}} \equiv \kappa T \sum_i \ln \frac{\sinh[\hbar(\omega_i - \Delta\omega_i)/2\kappa T]}{\sinh(\hbar\omega_i/2\kappa T)}, \tag{4.27}$$

where deformation-induced polarization of the electronic carrier shifts the i-th vibration frequency from ω_i to $\omega_i - \Delta\omega_i$. Here κ is the Boltzmann constant and T is the absolute temperature. At absolute zero the carrier-induced change of vibrations' free energy is just the net carrier-induced shift of vibrations' zero-point energy:

$$\Delta F_{vib} = -\frac{\hbar}{2}\sum_i \Delta\omega_i. \tag{4.28}$$

At finite temperatures the reduction of vibrations' free energy can be approximated by

$$\Delta F_{vib} \cong -\sum_i \frac{\hbar\Delta\omega_i}{2} \coth\left(\hbar\omega_i/2\kappa T\right), \tag{4.29}$$

when the carrier-induced changes of vibration frequencies are sufficiently small.

4.3 Scaling approach: large- and small-polaron formation

Adiabatic studies of self-trapping usually begin by passing to the limit of large atomic masses. Vibration frequencies then vanish as atoms become infinitely massive with the medium's stiffness remaining finite. Phenomena associated with atomic movements are thereby ignored. In particular, (1) carrier-induced softening of atoms' vibrations, (2) polaron motion and (3) non-adiabatic transitions are suppressed. In this limit a polaron's energy becomes just the adiabatic potential of Eq. (4.22). A self-trapped carrier's wavefunction is that which minimizes this energy.

A scaling analysis of the minimized adiabatic potential ground state yields its fundamental features without requiring explicit solution of the nonlinear differential equation, Eq. (4.19). The scaling analysis considers the effect on the energy of the adiabatic ground state of changing the length-scale of its electronic wavefunction (Emin and Holstein, 1976). In particular, r is replaced by r/L in $\phi_n(r)$ while its normalization is maintained by multiplying it by $L^{-d/2}$, where d is the electron's dimensionality. The energy functional obtained by this replacement, $E(L)$, is then studied as a function of the dimensionless scaling factor, L. Altering the length-scale of the wavefunction of the adiabatic ground state necessarily increases the corresponding energy. Therefore, the minimum of the energy functional occurs at $L = 1$.

Applying the scaling procedure to the adiabatic potential of Eq. (4.22) (plus changing integration variables) ultimately yields the energy functional:

$$E(L) = \frac{T_e}{L^2} - \frac{V_L}{L} - \frac{V_S}{L^d}, \tag{4.30}$$

where

$$T_e \equiv \frac{\hbar^2}{2m} \int dr |\nabla_r \phi_n(r)|^2, \tag{4.31}$$

$$V_L \equiv \frac{e^2}{2} \left(\frac{1}{\varepsilon_\infty} - \frac{1}{\varepsilon_0} \right) \int dr \int dr' \frac{|\phi_n(r)|^2 |\phi_n(r')|^2}{|r - r'|}, \tag{4.32}$$

and

$$V_S \equiv \left(\frac{F^2}{2k} \right) V_c \int dr |\phi_n(r)|^4. \tag{4.33}$$

As shown in Fig. 4.2, with only the long-range electron–phonon interaction the energy functional $E(L)$ has a single finite-energy minimum. This minimum occurs at $L = 2T_e/V_L$ where $E(L) = -V_L^2/4T_e$. Net binding occurs because the lowering of the carrier's energy, $T_e/L^2 - 2V_L/L = -3V_L^2/4T_e$, is 3/2 of the associated deformation energy, $V_L/L = V_L^2/2T_e$. The requirement that $L = 1$ at the minimum insures that $2T_e = V_L$ for the ground state.

A qualitatively different situation prevails with only the short-range electron–phonon interaction. As depicted in Fig. 4.3, the energy functional with full ($d = 3$) electronic dimensionality has minima at $L = 0$ and at $L = \infty$ separated by a maximum at $L = 3V_S/2T_e$ (Emin, 1973a). These two minima respectively correspond to a self-trapped carrier that collapses to a severely localized state and to a free carrier that extends over all space. The maximum, $8T_e^3/27V_S^2$, is termed the "barrier to self-trapping" because a free carrier must transcend this barrier in order to self-trap (Rashba, 1957; Toyozawa, 1961). As such, a free carrier experiences a time delay before self-trapping (Mott and Stoneham, 1977). In particular, *persistent photo-conductivity*

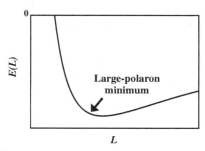

Fig. 4.2 The finite-radius of the minimum of the adiabatic energy functional $E(L)$ for a carrier in a three-dimensional ionic medium indicates that its self-trapped carrier forms a large polaron.

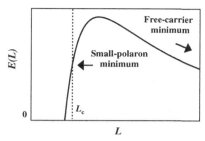

Fig. 4.3 No well-defined minimum exists for the adiabatic energy functional $E(L)$ of a carrier interacting with a three-dimensional continuum via a short-range electron–phonon interaction. A carrier either remains free, $E(L) \to 0$ as $L \to \infty$, or the carrier self-traps as a small polaron: $E(L) \to -\infty$ as $L \to 0$. The shrinking of a carrier is limited to L_c by a real material's atomicity. Small-polaron formation is indicated by $\partial E(L)/\partial L > 0$ at L_c.

occurs when a photo-carrier moves freely for a long time before it collapses into an energetically favorable severely localized state whose energy is significantly modified by the short-range electron–phonon interaction (Henry and Lang, 1977).

The three-dimensional continuum model with the short-range electron–phonon interaction gives $E(L) \to -\infty$ as $L \to 0$. This behavior describes a carrier becoming self-trapped at an energy that becomes infinitely deep as the carrier's potential well becomes infinitesimally small. However, beyond the continuum model in a real material this self-trapping *collapse* actually saturates as the size of the potential well shrinks to atomic dimensions. This saturation can be described by imposing a small-L cut-off on $E(L)$ at the value $L = L_c$. The value of $E(L_c) \equiv E_{sp}$ is the energy of a *small* polaron. By contrast, an unsaturated finite-radius polaron minimum, as found with the three-dimensional long-range electron–phonon interaction, indicates formation of a *large* polaron.

In the adiabatic limit, the small-polaron minimum at $E(L_c)$ coexists with the free-carrier minimum at $E(\infty)$. The small-polaron minimum is energetically stable if $E(L_c) < E(\infty) = 0$. In terms of microscopic parameters, a small-polaron's energetic stability requires that the lowering of its potential energy, $F^2/2k$, exceed the energy required to confine an electronic carrier to a site, essentially half the electronic bandwidth, $W/2 = 6t$ for a tight-binding band in a cubic crystal.

The dichotomy between a carrier remaining free and collapsing to form a small-polaron persists if the carrier's motion is limited to a plane ($d = 2$) with a short-range interaction. In particular, the solution is severely localized if $V_S > T_e$ and is delocalized if $V_S < T_e$. A type of two-dimensional self-trapping, electronically stimulated desorption from a surface, will be addressed in Chapter 16.

Self-trapping in quasi-one-dimensional materials has received considerable attention. Analysis of Eq. (4.30) reveals that a finite-radius solution occurs for a

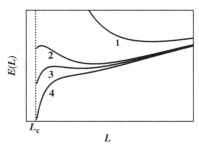

Fig. 4.4 The large-polaron minimum (curve 1) shrinks and falls in energy as the short-range interaction increases. As the short-range interaction strengthens (curve 2) a meta-stable small-polaron minimum appears at L_c. The small polaron is stabilized as the interaction is increased further (curve 3). Ultimately (curve 4) the large-polaron minimum disappears.

carrier whose motion is limited to an isolated chain ($d = 1$). The energy function then has its solitary minimum at $L = 2T_e/V_S$ where $E(L) = -V_S^2/4T_e$. Explicit solution for a one-dimensional chain yields a ground-state energy of $-t/3L_p^2$ for a self-trapped state whose half-length encompasses $L_p = 2t/(F^2/2k)$ chain units with an inter-unit electronic transfer energy of t (Holstein, 1959a). However, self-trapping in a quasi-one-dimensional anisotropic material requires severe localization in the two directions perpendicular to the chain. In particular, such localization costs about $2(2t_t)$, the half-bandwidth $2t_t$ in each of the two directions transverse to the chain. Thus the binding energy of a quasi-one-dimensional polaron is $-t/3L_p^2 + 4t_t$. Stability of a quasi-one-dimensional large polaron with respect to its carrier remaining free therefore requires very strong anisotropy of the electronic transfer energies: $t/t_t > 12L_p^2$ (Emin, 1986; Schüttler and Holstein, 1986).

Often a carrier interacts both with displacements of distant ions and with those of the atoms it contacts. The energy functional Eq. (4.30) describes both long- and short-range interactions. For a carrier confined to a chain ($d = 1$) embedded within an ionic medium the contributions to $E(L)$ from both interactions are simply summed as both are proportional to $1/L$. For a carrier confined to a plane ($d = 2$), the short-range interaction is proportional to $1/L^2$. As such, its effect is to simply replace the coefficient of the confinement-energy contribution T_e by $T_e - V_S$. With $d = 3$ long- and short-range electron–phonon interactions combine to produce the four possible polaron scenarios depicted in Fig. 4.4. As illustrated in curve 1, only a large polaron can form for a weak short-range interaction. Increasing the strength of the short-range interactions facilitates formation of a small polaron that may either be meta-stable or stable with respect to the large-polaron minimum, curves 2 and 3, respectively. As shown in curve 4, only small-polaron formation is possible with a sufficiently strong short-range interaction: $V_S > T_e^2/3V_L$. In particular, the long-range electron–phonon interaction facilitates a carrier's collapse into a small polaron.

4.4 Self-trapping beyond the adiabatic limit

Beyond the adiabatic limit (infinitely massive atoms) atoms can vibrate. Atomic vibrations enable both carrier-induced softening and strong-coupling polarons' motion. Atoms' vibrations affect polarons' energies, stability and their ability to form. In particular, the small-polaron and free-carrier coexistence domain contracts as atoms' vibration frequencies increase.

The potential experienced by a self-trapped carrier changes as atoms vibrate. The adjustment of a self-trapped carrier to this changing potential alters the stiffness constants and vibration frequencies associated with these atomic vibrations. If the carrier-induced shifts of atomic frequencies are sufficiently small the change of the free energy of atoms' vibrations is given by Eq. (4.27). For dispersion-less optic vibrations of frequency ω the change of vibrations' free-energy just depends on the net shift of atoms' vibration frequencies:

$$\Delta F_{\text{vib}} \cong -\frac{\hbar\coth(\hbar\omega/2\kappa T)}{2}\sum_i \Delta\omega_i. \tag{4.34}$$

Furthermore, the sum of the carrier-induced shifts of the optic frequencies is simply related to the sum of diagonal ($u = u'$) carrier-induced stiffness changes (Emin, 2000a):

$$\sum_i \Delta\omega_i = \frac{\omega}{S}\sum_{n'\neq n}\frac{\int dr'\phi_n^*(r')\psi_{n'}(r')\int dr\psi_{n'}^*(r)\phi_n(r)\int du\ Z(r'-u)\,Z(r-u)}{E_{n'}-W_n}$$

$$= 2\omega\sum_{n'\neq n}\frac{\int dr'\rho_{n,n'}^*(r')\int dr\rho_{n,n'}(r)I(r,r')}{E_{n'}-W_n}, \tag{4.35}$$

where Eqs. (4.15) and (4.23) have been utilized and the overlap density has been defined by

$$\rho_{n,n'}(r) \equiv \psi_{n'}^*(r)\phi_n(r). \tag{4.36}$$

Combining Eqs. (4.34) and (4.35) yields the carrier-induced change of the free energy of dispersion-less optic vibration for a self-trapped carrier in state n:

$$\Delta F_{\text{vib}} \cong -\hbar\omega\coth(\hbar\omega/2\kappa T)\sum_{n'\neq n}\frac{\int dr'\rho_{n,n'}^*(r')\int dr\rho_{n,n'}(r)I(r,r')}{E_{n'}-W_n}. \tag{4.37}$$

This carrier-induced shift of atomic-vibrations' free-energy contributes to the polaron energy and thereby affects its stabilization. In the zero-temperature limit this shift is just that from atoms' zero-point vibrations, proportional to $\hbar\omega$. The shift tends to be largest for very large self-trapped states as occur for molecular polarons (Emin, 2000a).

In the adiabatic limit, where atomic vibrations are suppressed, self-trapped carriers are strictly localized. Beyond the adiabatic limit, however, atoms vibrate and self-trapped carriers can move. Concomitantly, a strong-coupling polaron's mass falls from infinity in the adiabatic limit to a finite value that generally greatly exceeds that of a free electronic carrier. The associated polaron bandwidth rises from zero in the adiabatic limit ($\omega = 0$) to a value that is less than the characteristic phonon energy.

A large polaron's effective mass is a weighted measure of the masses of the ions whose shifts transfer the potential well containing the self-trapped carrier (Schüttler and Holstein, 1986; Emin and Hillery, 1989). Elements of the effective mass tensor can be written as the sum of the product of the masses of displaced ions' and their self-trapping strain fields in each of two principal directions (Emin and Hillery, 1989, p. 2388). In an isotropic medium a large polaron's scalar effective mass is often simply expressed in terms of the bare carrier mass m and the Fröhlich coupling constant, α of Eq. (3.3) by (Fröhlich, 1963):

$$m_{\mathrm{LP}} \approx m\alpha^4. \tag{4.38}$$

The strong-coupling condition $\alpha \gg 1$ ensures that the large-polaron effective mass greatly exceeds the bare carrier mass. Using the definition of α along with expressions for the binding energy $E_{\mathrm{LP}} = V_{\mathrm{L}}^2/4T_{\mathrm{e}}$ and spatial extent $R_{\mathrm{LP}} = (2T_{\mathrm{e}}/V_{\mathrm{L}})a$ of the large polaron, its effective mass can be rewritten as (Emin and Hillery, 1989; Emin, 1993):

$$m_{\mathrm{LP}} \approx \frac{E_{\mathrm{LP}}}{\omega^2 R_{\mathrm{LP}}^2}, \tag{4.39}$$

where $m_{\mathrm{LP}} \to \infty$ as $\omega \to 0$, the adiabatic limit. Knowing this mass enables estimation of the kinetic energy at which a large-polaron's de Broglie wavelength falls to its characteristic spatial extent $\sim R_{\mathrm{LP}}$:

$$E_{\mathrm{c}} \approx \frac{(\hbar/R_{\mathrm{LP}})^2}{m_{\mathrm{LP}}} \approx \hbar\omega \left(\frac{\hbar\omega}{E_{\mathrm{LP}}}\right). \tag{4.40}$$

Noting the self-trapping condition, $E_{\mathrm{LP}} > \hbar\omega$, it becomes evident that $E_{\mathrm{c}} < \hbar\omega$.

Small-polaron motion is qualitatively different from large-polaron motion. As illustrated in the two-site example of Section 3.3, an energy barrier is encountered as a small polaron moves between sites. Since the change of a carrier's electronic energy as it passes over a barrier exceeds its electronic transfer energy, a carrier loses coherence in such transfer processes. Coherent small-polaron transport is nonetheless possible with direct tunneling through this energy barrier. Such tunneling exchanges the pattern of atomic deformations about a site occupied by a small polaron with the atomic-deformation pattern surrounding an unoccupied site. These atomic tunneling processes are also described as "diagonal transitions" in which atomic displacements are exchanged between sites without phonons being emitted or absorbed (Holstein, 1959b, pp. 348–355). The width of a small-polaron band is a measure of the energy splitting produced by this tunneling. As such, small-polaron bandwidths vanish in both classical and adiabatic limits, $\hbar \to 0$ and $\omega \to 0$, since such tunneling is then suppressed.

Consider tunneling between the atomic deformation patterns of the two minima of the ground-state adiabatic potential Eq. (3.14) plotted in Fig. 4.5. The energy splitting of the adiabatic ground state is twice the matrix element of the Hamiltonian between local functions centered at different minima. Ignoring the t-dependent rounding of the barrier and approximating the mutually orthogonal local functions as displaced-harmonic-oscillator ground-state wavefunctions, the tunnel splitting is seen to be much less than the phonon energy $\hbar\omega$:

$$2\langle X\left(x \pm F/\sqrt{2k}\right)|H|X\left(x \mp F/\sqrt{2k}\right)\rangle \cong \hbar\omega\left[\sqrt{\frac{(F^2/2k)}{\hbar\omega}}\,e^{-(F^2/2k)/\hbar\omega}\right], \quad (4.41)$$

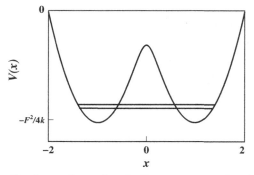

Fig. 4.5 With small-polaron formation the lowest energy levels of the two-site ground-state adiabatic potential reside at an energy of $\cong \hbar\omega/2$ above the minima at $\approx -F^2/4k$. Tunneling between these two minima splits these levels' degeneracy thereby forming an energy doublet which is represented by the figure's two horizontal lines.

where $F^2/2k \gg \hbar\omega$. The ground-state tunnel spitting becomes even smaller, $\approx 2t \exp[-(F^2/2k)/(\hbar\omega)]$, if the electronic transfer energy is small enough to invalidate the adiabatic approach, $2t \ll [(F^2/2k)(\hbar\omega)]^{1/2}$ (Holstein, 1959b, pp. 370–375).

Overlap between displaced harmonic oscillators produces the exponential reduction factor of the small-polaron tunnel splitting. Increasing the energies of these harmonic oscillators' vibrations increases the spatial extents of their wavefunctions. However, the wiggles of the wavefunction of a harmonic oscillator increase with its energy. As a result the net overlap between displaced harmonic oscillators decreases as their energies are increased. Thus the thermally averaged overlap between displaced oscillators falls with increasing temperature. In particular, the exponential overlap factor is reduced to $\exp\{-[(F^2/2k)/\hbar\omega]\coth(\hbar\omega/2\kappa T)\}$ at finite temperatures.

Small-polaron bands are typically too narrow to support coherent motion. For example, the adiabatic tunnel splitting of Eq. (4.41) is < 0.01 eV for even very modest values of the parameters: $\hbar\omega = 0.1$ eV and $(F^2/2k)/\hbar\omega = 3$. For $\hbar\omega = 0.03$ eV and $(F^2/2k)/\hbar\omega = 10$ the adiabatic tunnel splitting is $< 10^{-5}$ eV. Thus, the site-to-site disorder energy produced by charge centers, defects and even applied fields often exceeds the small-polaron bandwidth. Coherent motion is then destroyed and small polarons move incoherently via thermally assisted hopping as described in Chapter 11.

Variational studies indicate that a carrier added to a three-dimensional deformable crystal either remains nearly free or collapses to form a small polaron (Toyozawa, 1961; Emin, 1973a). As depicted in the uppermost panel of Fig. 4.6, consistent with the results of Section 4.3 for the adiabatic limit ($\omega = 0$), the zero-momentum ground states of these two solutions coexist. The free carrier is energetically stable when $F^2/2k < W/2$ and the small-polaron solution is energetically stable when $F^2/2k > W/2$. As shown in the middle panel of Fig. 4.6, beyond the adiabatic limit, as $\hbar\omega/(W/2)$ rises from zero, the values of $(F^2/2k)/(W/2)$ between which these two solutions coexist shrinks. In particular, a small polaron cannot form for too weak an electron–phonon coupling and a carrier cannot exist as nearly free for too strong an electron–phonon coupling. The energy of the small-polaron's zero-momentum state is given by

$$E_{\mathrm{SP}} = -z\hbar\omega\sqrt{\left(\frac{F^2/2k}{\hbar\omega}\right)}\exp\left(-\frac{F^2/2k}{\hbar\omega}\right) - F^2/2k, \qquad (4.42)$$

where z is the number of nearest neighbors ($z = 6$ for a cubic lattice) and small-polaron transfer is taken as adiabatic. The free-carrier energy is approximately that obtained by treating the electron–phonon interaction as a small perturbation:

$$E_{FC} = -W/2 - \frac{(F^2/2k)\hbar\omega}{W/2}.$$ (4.43)

As depicted in the lowest panel of Fig. 4.6, the coexistence regime disappears when the electronic bandwidth divided by the phonon energy is small enough. Then these variational treatments fail between their free-carrier and small-polaron limits.

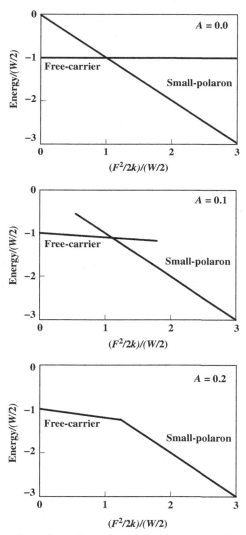

Fig. 4.6 The free-carrier and small-polaron ground-state energies in units of half the electronic bandwidth $W/2$ are plotted against the electron–phonon coupling strength $(F^2/2k)/(W/2)$ for three values of $A \equiv \hbar\omega/(W/2)$. The coupling strengths between which both solutions coexist diminishes as A increases from 0, the adiabatic limit.

To understand the condition limiting small-polaron formation, consider the energy functional for the energy of a zero-momentum small-polaron divided by $F^2/2k$:

$$\frac{E(y)}{F^2/2k} = -z\frac{\exp\left(-\dfrac{F^2/2k}{\hbar\omega}y^2\right)}{\sqrt{\left(\dfrac{F^2/2k}{\hbar\omega}\right)}} - 2y + y^2, \tag{4.44}$$

where y is a dimensionless measure of the displacements of the equilibrium positions of atoms surrounding the carrier. The first contribution on the right-hand side of Eq. (4.44) is proportional to the lowering of the small-polaron's energy due to its adiabatic tunneling to adjacent sites. The second and third contributions on the right-hand side of Eq. (4.44) are proportional to the lowering of the carrier's energy and the associated increase of the solid's strain energy, respectively. With small-polaron formation this functional has a minimum near $y = 1$. However, as shown in Fig. 4.7, this small-polaron minimum disappears when the coupling strength $(F^2/2k)/\hbar\omega$ becomes small enough for the tunneling contribution to the small-polaron energy to become significant.

The electron–phonon coupling strength becomes too strong for a carrier to move freely with its interaction with the phonons being just a perturbation when the two contributions to E_{FC} of Eq. (4.43) become comparable to one another. This condition can be expressed as:

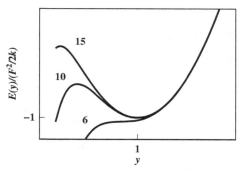

Fig. 4.7 The energy functional $E(y)$ in units of $F^2/2k$ is plotted against the deformation parameter y for three values of the electron–phonon coupling parameter $(F^2/2k)/\hbar\omega$. The minimum of $E(y)$ at $y \approx 1$, indicating small-polaron formation, vanishes when $(F^2/2k)/\hbar\omega$ falls below 6.

$$F\left(\frac{\hbar}{W/2}\right) > \sqrt{M\hbar\omega}. \tag{4.45}$$

Thus, the carrier can no longer move freely if the product of the force it exerts on the surrounding atoms F multiplied by the time that a free carrier can be held at a site $2\hbar/W$ exceeds atoms' zero-point momentum $\sqrt{M\hbar\omega}$ (Emin, 1973a). That is, a carrier can no longer move freely if it lingers at a site long enough for its presence to significantly alter atoms' motion.

So far this discussion has only considered a carrier interacting with zero-point phonons. However, the variational calculations and the physical arguments governing the validity of the small-polaron and free-carrier solutions survive extension to finite temperature (Emin, 1972; Emin 1973a). As the temperature is raised the domain in which small-polaron and free-carrier solutions coexist widens. This situation suggests a type of thermally driven conductivity transition. In particular, below the transition temperature all carriers self-trap and have low mobility while above the transition temperature conduction is dominated by high-mobility free carriers (Emin, 1972; Emin 1973a).

4.5 Disorder-assisted small-polaron formation

Evidence of small polarons is found among non-crystalline solids. In these instances the absence of crystallinity presents charge carriers with a disordered environment. By itself, disorder fosters electronic localization and makes extended-state motion tortuous (Anderson, 1958). Since confining and slowing a charge carrier facilitates the displacing of surrounding atoms, disorder aids a carrier's collapse into a small polaron. By driving a moderately localized or tenuously moving carrier to collapse into a severely localized self-trapped state, electron–phonon interactions amplify the localizing tendencies of disorder. For example, time-of-flight measurements indicate that the high-mobility quasi-free motion of holes in crystalline orthorhombic sulfur collapses to low-mobility small-polaron hopping motion as this molecular crystal is heated into its "softening regime" where the S_8 molecules relinquish the regularity of their locations with respect to one another (Ghosh and Spear, 1968).

The electron–phonon interaction deepens the potential well of a localized carrier. The deepening of the carrier's potential well in turn increases the severity of the carrier's localization. This regenerative nature of self-trapping is embodied in the nonlinearity of the differential equation governing self-trapping, Eq. (4.19). As a result, sufficiently localized states in covalent glasses are expected to collapse into small polarons (Anderson, 1972).

The scaling treatment of self-trapping in the adiabatic limit of Section 4.3 can be employed to show how increasing confinement causes a localized carrier to collapse into a small polaron (Emin and Holstein, 1976). The localized state's energy functional is analogous to Eq. (4.30):

$$E(L) = \frac{T_e}{L^2} - \frac{V_S}{L^3} - \frac{V_D}{L^s},$$

(4.46)

where L is the length-scaling factor and the energetic constants T_e and V_S are defined by Eqs. (4.31) and (4.33). The disorder constant associated with the localized state's bare electronic potential, $-V_0/r^s$, is given by:

$$V_D \equiv \int dr |\phi_n(r)|^2 \left(\frac{V_0}{r^s} \right).$$

(4.47)

The energy functional of Eq. (4.46) is plotted in Fig. 4.8 for $s = 1$. The parameters are chosen such that the carrier is delocalized in the absence of the localizing potential, $V_D = 0$. Increasing the strength of the localizing potential introduces a meta-stable small-polaron state. Further strengthening of the localizing potential stabilizes the small-polaron state. A still stronger localizing potential eliminates all but the small-polaron solution. Thus, sufficiently localized carriers are converted into small polarons.

In a crystal the electronic eigenstates are extended with uniform charge density. Imposing disorder introduces localized states while rendering the charge densities of extended states non-uniform (Mott and Davis, 1979). The energy separating an energy band's localized states from its extended states is termed its mobility edge. The Emin–Holstein scaling analysis has been used to argue that the electron–phonon

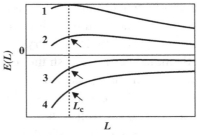

Fig. 4.8 Imposition of a localizing potential can induce a free carrier in a uniform covalent medium (curve 1) to form a small-polaron (arrows). As a localizing potential is strengthened a meta-stable small-polaron is introduced (curve 2) and then stabilized (curve 3). A strong enough potential triggers a carrier's collapse into a localized small polaron (curve 4).

interaction causes sufficiently non-uniform extended states above the mobility edge to behave analogously to localized states by collapsing into small polarons (Cohen, Economou and Soukoulis, 1983).

Another approach to disorder-induced collapse of extended-state carriers into small polarons extends the adiabatic theory of small-polaron formation in a crystal to a covalent glass (Emin and Bussac, 1994). A stable carrier in a crystal will remain free or form a small polaron depending on the magnitude of the electronic band-width W. Within the tight-binding model the electronic half-bandwidth is the product of the number of nearest neighbors z and the magnitude of the electronic transfer energy t: $W/2 = zt$. If $zt > F^2/2k$ a carrier moves freely between sites without polaron deformation. Alternatively, for $zt < F^2/2k$ the carrier forms a small polaron. A covalent glass is characterized by variations of its bond angles. These differences introduce significant dispersion of its electronic transfer energies. This local dispersion of electronic transfer energies inhibits a carrier's ability to avoid forming a small polaron. In particular, a carrier can only avoid forming a small polaron at site g if its net electronic transfer energy from that site to neighboring sites $g + h$ is large enough:

$$\sum_{h=1}^{z'} t_{g,g+h} > \frac{F^2}{2k},\qquad(4.48)$$

where z' indicates that the summation excludes transfers to sites at which small-polaron formation occurs. Indeed, such transfers are forbidden in the adiabatic limit $\hbar\omega \to 0$. This feature constrains a free carrier to avoid sites at which it would collapse into a small polaron. This restriction on a free carrier's movement also fosters its collapse into a small polaron. As a result of this feedback effect modest transfer-energy disorder can significantly reduce the electron–phonon coupling strength required for global small-polaron formation (Emin and Bussac, 1994). Indeed steady-state transport studies in covalent glasses indicate small-polaron hopping (Emin, Seager and Quinn, 1972; Seager, Emin and Quinn, 1973; Seager and Quinn, 1975; Klaffe and Wood, 1976; Lewis, 1976a; Lewis, 1976b; Moustakas and Paul, 1977; Baily and Emin, 2006a; Baily, Emin and Li, 2006b).

4.6 Synopsis

An electronic carrier typically adjusts to the sluggish motions of the relatively heavy atoms that surround it. As such, a carrier may find itself bound within the potential well produced by suitable atomic displacements. This atomic configuration will appear nearly static to the bound carrier if its binding energy is larger than the characteristic phonon energy. The carrier is then self-trapped, bound within a potential well produced by surrounding atoms being displaced to new equilibrium positions consistent with the carrier's presence.

A polaron denotes the unit comprising the self-trapped carrier and the surrounding atoms at displaced equilibrium positions. The spatial extent of a large polaron's self-trapped carrier increases continuously with the magnitude of the electronic bandwidth and typically extends over multiple atomic sites. By contrast, a small polaron's self-trapped carrier collapses to the smallest relevant structural unit.

A self-trapped carrier interacting with the displaceable ions of a three-dimensional polar medium only via the long-range (Fröhlich) electron–phonon interaction will generally form a large polaron. By contrast, a carrier with only a short-range electron–phonon interaction will either remain free or collapse into a small polaron. A small polaron will be energetically stable if its binding energy $F^2/2k$ exceeds the electronic band's half-width $W/2$. Exceptionally, a short-range electron–phonon interaction supports a large polaron in a quasi-one-dimensional material if its electronic transfer energies are extremely anisotropic.

Typically a large polaron can form in an ionic solid. By contrast, by itself the short-range electron–phonon interaction is often neither strong enough nor the electronic energy bands narrow enough to produce an energetically stable small polaron $F^2/2k > W/2$. Nonetheless, adding the localizing effects of disorder and/or the long-range electron–phonon interaction frequently triggers a carrier's collapse into a small polaron.

Self-trapping is often studied within the adiabatic limit where atoms' motion is arbitrarily slow. Beyond the adiabatic limit atoms are permitted to vibrate with finite frequencies. Moreover, a self-trapped carrier's adjusting to the changing positions of atoms shifts their vibration frequencies. Atomic vibrations also enable a polaron to move and thereby garner a finite bandwidth. The width of a large-polaron band is less than the characteristic phonon energy. A small-polaron band is usually much narrower than even a large-polaron band. A large polaron moves coherently with its motion impeded by occasional scattering events. Small-polaron motion is generally incoherent via a succession of phonon-assisted hops.

Beyond the adiabatic limit a small polaron can only exist beyond a minimum value of the short-range electron–phonon coupling strength. A free carrier can only exist below a maximum value of the short-range electron–phonon coupling strength. With $W \gg \hbar\omega$ there is a span of short-range electron–phonon coupling strengths within which free carriers and small polarons can coexist. This coexistence domain increases with increasing temperature. Thus, if only small polarons form at low temperatures above a transition temperature some carriers will abruptly be freed. Below this transition temperature transport will be by low-mobility small-polaron hopping. Above the transition temperature transport may be dominated by conventional higher mobility carriers.

5

Dopant- and defect-related small polarons

Small polarons' associations with dopants and defects are often more complex than those between carriers and dopants in conventional semiconductors. To illustrate these associations, three examples of dopant- and defect-related small polarons are succinctly described. In the simplest situation a small polaron occupies one of the equivalent sites adjacent to a dopant. A more complex case envisions a large polaron being induced to collapse into a small polaron when it is confined to the vicinity of a dopant. The final scenario envisions one of two electrons bound to a defect collapsing to a small polaron as the other electron is liberated.

Dopants in semiconductors whose carriers form small polarons can be more intricately involved in transport than are dopants in conventional high-mobility semiconductors. In particular, electronic transport in conventional high-mobility semiconductors is governed by very low concentrations of carriers and dopants. By contrast, transport in small-polaron semiconductors typically involves very high concentrations of their low-mobility carriers and the dopants which spawn them. As a result direct hops between dopant-related states can dominate transport even well above room temperature. To illustrate these situations an example in which such impurity conduction might prevail is discussed.

5.1 Small polarons bound to dopants and defects

Hole-type charge carriers are introduced in crystalline MnO by replacing a small concentration of Mn^{2+} cations with Li^{1+} cations. Measurements of the dc charge transport (conductivity, Seebeck coefficient, and Hall mobility) indicate that liberated hole-like carriers form small polarons and move by thermally assisted hopping between Mn sites (Crevecoeur and Wit, 1970). Since each liberated carrier is severely localized as a small polaron, a carrier bound to the Li dopant is similarly localized. In particular, a carrier bound to a Li dopant forms a Mn^{3+} cation at one of the 12 nearly equivalent next-nearest neighbor Mn sites surrounding the Li cation (Bosman and van Daal, 1970). In other words, the dopant state is just a small-polaron bound to a dopant. By contrast, a conventional dopant state, such as the donor state formed by substituting a phosphorus atom for one of silicon's atoms, encompasses very many atoms.

A mixed situation has been reported for holes that are introduced in NiO by substituting Li for Ni (Bosman and van Daal, 1970). DC electronic transport measurements have been interpreted as indicating that liberated carriers move as large polarons. By contrast, these authors envision a hole-like carrier bound to a Li dopant forming a small polaron which resides on one of the 12 Ni next-nearest-neighbor sites that surround the dopant. Hopping of a bound small polaron about its Li dopant contributes to the doped materials' ac conductivity. The large-polaron holes in NiO are ascribed to the wide oxygen p-band produced by large (~ 1.4 Å) strongly overlapping oxygen anions. By contrast, NiO's small-polaron holes are attributed to the relatively narrow nickel d-bands spawned by weak overlap between small (~ 0.7 Å) nickel cations. Only when a hole is adjacent to a Li-dopant is it stabilized as a nickel-based small-polaron rather than as an oxygen-based large-polaron.

Removing an atom from a material, i.e. replacing the atom with a vacancy, can also introduce charge carriers. In particular, oxygen vacancies are common defects in oxides. In these ionic materials oxygen ions are generally assigned a valence of 2−. However, an isolated oxygen atom can only bind a single additional electron. The positive charges of adjacent cations must assist in binding a second electron near an oxygen atom to produce its formal valence of 2−. For example, each oxygen site in $SrTiO_3$ has two Ti^{4+} nearest neighbors and four Sr^{2+} next-nearest neighbors. In these instances, electrons that are attracted to an oxygen atom and its adjacent cations may remain attracted, albeit more weakly, when the oxygen atom is removed. Thus, an oxygen atom and the corresponding vacancy may both formally bind two electrons.

In principle, two electrons can be liberated from a doubly occupied oxygen vacancy. Moreover, there are two complementary scenarios for thermally induced release of a singlet pair of electrons bound to an oxygen vacancy. In the first circumstance, as with bipolaron formation (discussed in Chapter 7), the energy with which the electron–phonon interaction indirectly binds the vacancy's two electrons together overwhelms their mutual Coulomb repulsion. The energy needed to liberate the first electron is then greater than that to free the second electron. Thus thermal de-trapping of one electron of the pair portends relatively rapid thermal release of the second electron. In the second scenario the two electrons' mutual Coulomb repulsion energy predominates. It then takes much less energy to liberate the first electron than to free the second electron. Freeing just one electron of a singlet pair localized at an oxygen vacancy not only produces the liberated carrier but concurrently generates a localized spin and optical absorption center associated with the residual bound electron. That is, as one of the pair of localized electrons is freed the remaining electron collapses into a bound small polaron.

Much larger densities of dopants and defects are usually required to significantly alter the electrical conductivity of a small-polaron semiconductor than that of a conventional semiconductor. In particular, a relatively high density of low-mobility small polarons is needed to produce electrical conductivity comparable to that generated by a low density of a conventional semiconductor's high-mobility carriers. The electrical conductivity of a small-polaron material is then insensitive to changes of its carrier density that would greatly affect a conventional semiconductor. Alternatively stated, a small-polaron material's Fermi energy appears "pinned".

5.2 Small-polaronic impurity conduction

Dopants provide charge carriers that can be liberated to move through a host material. Alternatively, *impurity conduction* envisions charge carriers moving directly between dopants. In conventional lightly doped semiconductors (e.g. phosphorus-doped silicon), freed carriers move through the host material with a much greater mobility than that for inter-dopant motion. Thus, impurity conduction only prevails when the temperature is low enough to preclude carriers being freed from their dopants. For example, impurity conduction in conventional semiconductors is only seen at cryogenic temperatures.

A different situation often occurs when charge carriers form small polarons. In these situations the small-polaron material's dopant density is usually orders of magnitude greater than that of common semiconductors. Then dopants are close to one another and charge carriers are never far from dopants. In the terminology of conventional semiconductors, one might say that impurity conduction then dominates charge transport.

To illustrate such situations consider small polarons in heavily doped $LaMnO_3$. The nominal (distortion free) $LaMnO_3$ crystal structure has Mn^{3+} ions at the eight corners of the cubic unit cell, O^{2-} ions centered at the middle of each of the cube's 12 edges, and a La^{3+} ion at the cube's center. Replacing a La^{3+} ion with a divalent ion (Ca^{2+}, Sr^{2+} or Ba^{2+}) introduces a hole. Assuming a random replacement of divalent cations for La^{3+}, Fig. 5.1 shows the probability that a dopant has n of the six nearest La cations also being replaced with dopants. The $n = 0$ curve shows that the probability of finding a dopant without another dopant at a nearest-neighbor dopant position falls rapidly with increasing doping, x. Much attention has focused on $0.2 < x < 0.5$, where this material becomes ferromagnetic. Even at

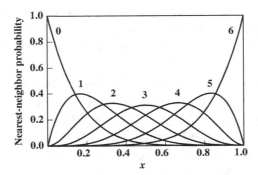

Fig. 5.1 The probability that random substitutions of dopants for La in $LaMnO_3$ leaves a dopant with n dopant neighbors is plotted against the doping fraction x for $0 \le n \le 6$.

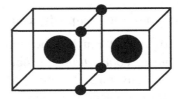

Fig. 5.2 Four Mn sites (small black circles) benefit most from the attractive Coulombic potentials provided by dopants (large black circles) in the centers of two adjacent unit cells of cubic $LaMnO_3$.

the lowest doping level in this range 75% of dopants have one or more dopants at nearest-neighbor La sites. That is, dopants are not widely dispersed.

The commonly envisioned small polaron in $LaMnO_3$ is a hole on a Mn site. The probability of a Mn site not having a dopant occupying at least one of the eight closest La sites is $(1 - x)^8$ (Miller *et al.*, 1961). With $x > 0.2$ this probability is very small, $< (0.8)^8 \approx 0.2$. Thus, a small polaron on a Mn site is generally close to dopants to which it is attracted.

When located between neighboring dopants a Mn site enjoys the sum of their attractive potentials. As illustrated in Fig. 5.2, the potential energy at a Mn cation will be especially low at the four sites that lie between two nearest-neighbor La-site dopants. In fact a magnetic susceptibility study of paramagnetic heavily doped $LaMnO_3$ concludes that its holes each occupy a localized molecular orbital comprising four Mn cations that align as a ferromagnetic to give a net spin of $4 \times 2 - 1/2 = 15/2$ (Tanaka *et al.*, 1983a). Figure 5.3 depicts the energies of such a cluster's four molecular orbitals. Other locations, such as the two Mn sites comprising the edge shared by three cubic unit cells inhabited by dopants, also

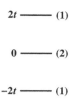

$2t$ ——— (1)

0 ——— (2)

$-2t$ ——— (1)

Fig. 5.3 The molecular energy levels of four Mn cations arranged in a square are split by their nearest-neighbor electronic transfer energy t. The bracketed numbers adjacent to the energy levels indicate the molecular orbitals' degeneracy.

have low potential energies. Holes presumably occupy such low-energy states with multiple holes occupying some clusters of contiguous dopants. In this way carriers in their ground-state distribute themselves in response to the disordered environment generated by the dopants.

Transport measurements in these materials' paramagnetic regime indicate thermally assisted multi-phonon hopping. In particular, analyses of electrical conductivity and Seebeck-coefficient measurements imply that the mobility is low and thermally activated (Heikes and Ure, 1961; Tanaka *et al.*, 1982; Palstra *et al.*, 1997). To obtain the mobility the thermally activated conductivity is presumed to be expressible as $\sigma = nq\mu$, the product of the carriers' density n, charge q and mobility μ. As explained in Section 12.2, the Seebeck coefficient is approximated as $\alpha = (\kappa/q)\ln[(N - n)/n]$, where N is the density of thermally accessible sites. A thermally activated mobility is inferred by comparing the measured temperature dependence of the conductivity with that of n deduced from the measured Seebeck coefficient.

Seebeck coefficients of heavily doped $LaMnO_3$ suggest that the phonon-assisted hopping of its holes occurs between a relatively small number of defect-related states rather than among the totality of Mn cations. The magnitudes of the high-temperature Seebeck coefficients are often much too small to just represent the doping concentration divided by the total concentration of Mn sites (Hundley and Neumeier, 1997). For example, the measured room-temperature values of the Seebeck coefficient are $\alpha \sim 5$ μV/K at 20% doping (Volger, 1954; Tanaka *et al.*, 1983b) while the Seebeck formula gives $\alpha = 120$ μV/K with $n/N = 0.2$. Moreover, the Seebeck coefficients and electrical conductivities are very sensitive to the introduction of modest densities of oxygen vacancies (Tanaka *et al.*, 1982; Aselage *et al.*, 2003). Transport of Mn-related compensated holes then appears to involve only a subset of Mn sites, e.g. those between adjacent dopants.

The magnitude and temperature dependence of the Hall mobility also indicate phonon-assisted hopping. Early measurements of heavily doped $LaMnO_3$ also revealed an anomalous signed (*n*-type for *p*-type doping) Hall mobility that was too small ($\ll 1$ cm^2/V-sec) for conventional transport (Volger, 1954). Subsequent Hall mobility measurements found it anomalously signed, very low and thermally activated with its activation energy well below that deduced for the drift mobility (Jaime *et al.*, 1997). The magnitude and temperature dependence of the Hall mobility is consistent with that for adiabatic small-polaron hopping (Emin and Holstein, 1969). Indeed, the low-mobility thermally activated charge transport deduced for heavily doped LaMnO$_3$ has often been attributed to small-polaron hopping between transition-metal ions (Volger, 1954; Miller *et al.*, 1961; Kertesz *et al.*, 1982; Tanaka *et al.*, 1982; Palstra *et al.*, 1997; Jaime *et al.*, 1997).

The *n*-type Hall effect observed in heavily doped *p*-type LaMnO$_3$ suggests hopping among a disordered arrangement of dopant-related localized states. Nearest-neighbor oxygen-mediated small-polaron hopping between Mn sites in LaMnO$_3$ would occur along its unit-cell's cube edges. However, no Hall-effect sign anomaly should occur for such hopping (Emin, 1971c). In fact, no Hall-effect sign anomaly is found in lightly doped cubic MnO where oxygen-mediated small-polaron hopping also presumably occurs between Mn sites along unit-cell cube edges (Crevecoeur and de Wit, 1970). Rather, Hall-effect sign anomalies are predicted for hopping that is not constrained to be along a cube's edges (Emin, 1977a). Indeed, Hall-effect sign anomalies are frequently observed for hopping conduction in non-crystalline solids.

In summary, electronic transport in the paramagnetic regime of heavily doped LaMnO$_3$ proceeds via multi-phonon hopping. Although uncertain, the localized states between which carriers hop seem to be dopant-related. An example of such a state is the four-Mn square unit subtended by a pair of dopants depicted in Fig. 5.2. Electronic transport would then constitute a type of small-polaronic impurity conduction.

6

Molecular polarons

6.1 General features

A molecular polaron is defined as a polaron whose self-trapped electronic charge carrier is primarily confined within a single molecule. The molecule may be a constituent of a molecular solid, a solid comprised of molecules, e.g. pentacene. The molecule might also be an additive as occurs with molecularly doped polymers. Often, as in biological materials, the embedding medium contains displaceable ions and polar molecules.

The concept of a molecular polaron emerges naturally when a carrier's slow intermolecular motion fosters its self-trapping on a molecule. In particular, the difficulty of intermolecular motion is associated with modest intermolecular electronic overlap and correspondingly narrow electronic energy bands. By contrast, electronic transfer among at least several of a molecule's atoms may be rapid.

One can envision three types of molecular polaron. As schematically depicted in Fig. 6.1a, these three kinds of molecular polaron are distinguished by the spatial extent of their electronic carrier. The largest carrier extends over all of the equivalent structural units of the molecule. These structural units can be individual atoms, bonds or multi-atomic chemical units such as rings of atoms. A smaller carrier is limited to a non-minimal subset of the equivalent structural units. The smallest type of molecular polaron has its carrier collapsed to a single structural unit.

The three types of molecular polaron are analogous to the three solutions to the nonlinear eigenvalue equation Eq. (4.19) which governs an electronic carrier's self-trapping in a deformable continuum. These three solutions are those for a free carrier, a large polaron and a small polaron.

The scaling procedure of Section 4.3 provides an efficient means of establishing the three kinds of molecular polaron. In particular, this method is used to find the minima of the adiabatic potential for an electronic carrier occupying a deformable molecule embedded within a medium containing displaceable ions and/or polar molecules (Emin, 2000a). The electronic kinetic energy term drives the electronic

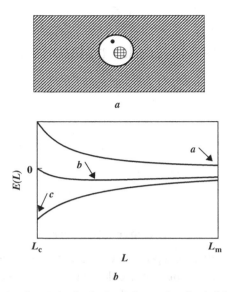

Fig. 6.1 Panel *a* is a schematic depiction of a molecule (white circle) embedded within a medium (diagonal-hatched area). Three types of molecular polaron are possible. (1) A molecular polaron's carrier encompasses all of the molecule's structures that feasibly can be occupied. (2) The carrier can occupy a non-minimal subset of these structures (square-hatched region within the molecule). (3) The self-trapped electronic carrier can collapse to a single structural unit (black dot). Panel *b* plots a molecular polaron's energy function $E(L)$ against its size parameter L for three values of the functional's physical parameters. The value L_m corresponds to the carrier's maximal expansion. The minimum value L_c indicates a carrier's collapse to its minimal size. The minimum value of $E(L)$ on each of the three curves is indicated with an arrow. Curve *a* depicts a maximally expanded carrier. Curve *b* corresponds to a carrier with intermediate expansion. Curve *c* indicates a collapsed carrier.

carrier to expand over the entire molecule. The electron–phonon interactions foster a carrier's localization. Here the short-range electron–phonon interaction describes the dependence of the energy of the carrier on displacements of the molecule's atoms that it contacts. The long-range Fröhlich electron–phonon interaction depicts the dependence of the carrier's potential energy on displacements of ions and polar molecules of the surrounding medium. The energy functional expressing the minimum of the adiabatic potential as a function of L, the scaling parameter characterizing the self-trapped carrier's spatial extent, is

$$E(L) = \frac{T_e}{L^2} - \frac{V_S}{L^d} - \frac{V_L}{L},$$
(6.1)

where d is the dimensionality of the carrier's expansion.

The three types of minima of $E(L)$ with respect to L shown in Fig. 6.1b correspond to three kinds of molecular polaron. As illustrated by curve a, when the electronic kinetic energy is large enough the minimum of $E(L)$ occurs at L_m, the maximum physically acceptable value of L. In this instance the carrier extends over the entire molecule. This solution is like the free-carrier solution found in Chapter 4 except that the carrier's spatial extent is limited by the boundary of the molecule. As depicted in curve b, with a sufficiently strong long-range electron–phonon interaction the minimum of $E(L)$ falls at a value of L lying between that corresponding to the electronic carrier collapsing to a single site L_c and L_m. Then the electronic carrier will occupy a subset of the molecule's equivalent sites. This solution is analogous to that for a large polaron. Curve c depicts the situation in which the molecule's carrier has a very small electronic kinetic energy. The minimum of $E(L)$ is then at L_c. The electronic carrier then collapses to a single structural subunit of the molecule. This solution is analogous to that for a small polaron.

As with polaron formation generally, the formation of a molecular polaron is associated with carrier-induced shifts of atoms' equilibrium positions and their vibration frequencies. As described in Section 4.2, carrier-induced alterations of atomic vibration frequencies occur as the electronic carrier adjusts to atoms' vibratory movements. As noted below Eq. (4.18) this effect increases strongly with the localized electron's spatial extent. Therefore carrier-induced shifts of vibration frequencies tend to be especially large for molecular polarons. Electronic carriers associated with intermediate-sized molecular polarons may be most polarizable since they are neither constrained by the carrier's collapse nor its molecule's spatial extent.

Carrier-induced shifts of atoms' vibration frequencies alter atoms' vibration entropy. Carrier-induced shifts of atoms' vibration entropy affect the rates with which carriers' execute phonon-assisted hops and the entropy transported with these hops. These features of polaron hopping provide an explanation of the empirically observed Meyer–Neldel compensation effect and the enhancement of carriers' Seebeck coefficients. These effects are discussed in Chapter 11 and Chapter 12, respectively.

The activation energies of polaron jump rates generally rise with jump distance. By itself, this effect fosters carriers making short hops. Applying an electric field promotes carriers making especially long jumps in concert with it. These two effects compete with one another. For a polaron formed through its carrier's long-range Coulomb-based electron–phonon interaction with ions and polar molecules surrounding it, the trade-off can produce a contribution to a polaron's hopping activation energy which falls in proportion to the square root of the electric-field strength. This effect is termed Frenkel–Poole behavior. It is also addressed in Chapter 11.

Holstein introduced the molecular-crystal model (MCM) as a simple pedagogic construct to address some aspects of polaron formation and small-polaron hopping

Fig. 6.2 A transition-metal cation (e.g. Mn in MnO) depicted as a black dot surrounded by six displaceable oxygen ligands (open circles) is represented in the molecular-crystal model as a diatomic molecule whose solitary deformation coordinate x is the deviation of its interatomic separation from its carrier-free equilibrium value a.

(Holstein, 1959a; 1959b). However, this model does not represent a real molecular solid. Its pseudo-molecules do not permit carrier-induced softening. It also ignores the long-range electron–phonon interaction characteristic of an electronic carrier in a polar medium. Indeed, the term polaron was coined to emphasize the importance of these interactions to self-trapping phenomena in polar and ionic media. As noted above, these features can generate the Meyer–Neldel and Frenkel–Poole effects.

The molecular-crystal model (MCM) envisions a regular array of pseudo-molecules that each has but one scalar deformation coordinate (Holstein, 1959a). In particular, the deformation coordinate of a pseudo-molecule at position g is denoted as x_g. The energy of a carrier on this molecule is taken to be $\varepsilon_g = -Fx_g$. Thus, adding a carrier to the molecule shifts the equilibrium value of its deformation coordinate without affecting its vibration frequency. This model was developed to provide an idealized depiction of self-trapping in a simple inorganic solid. For example, as illustrated in Fig. 6.2, an oversimplified model for the small-polaron hole in MnO presumes that the energy of a severely localized carrier on a transition-metal cation only depends on the positions of its nearest-neighbor oxygen ligands. The small polaron forms as the ligands undergo symmetry-breaking (Jahn–Teller) displacements about the Mn^{3+} hole thereby lifting its electronic degeneracy. The MCM represents this carrier-induced displacement of oxygen ions' equilibrium positions as that induced by a carrier occupying a non-degenerate electronic state on its pseudo-molecule's single scalar deformation coordinate.

6.2 Examples

Polarons formed on asymmetric molecules are widely studied. In particular, because of electro-photographic applications considerable effort has been devoted to studying polaron motion between asymmetric molecules embedded among polymers. For example, holes are thought to hop between p-diethylaminobenzaldehyde diphenylhydrazone (DEH) molecules dispersed among polycarbonate and polystyrene polymers

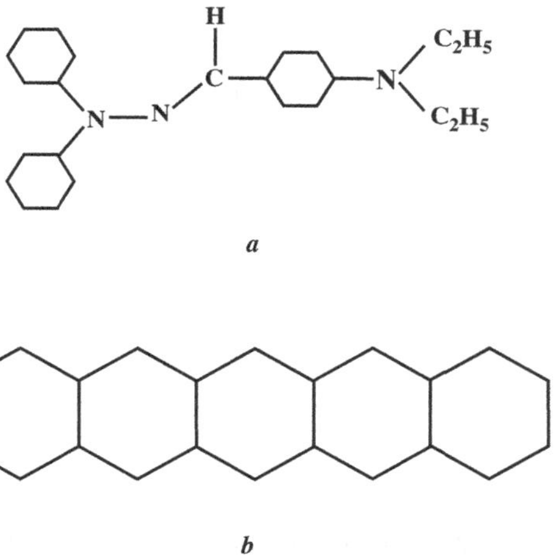

a

b

Fig. 6.3 Panel *a* depicts a DEH molecule. Panel *b* illustrates a pentacene molecule.

(Schein and Borsenberger, 1993). As illustrated in Fig. 6.3a, a DEH molecule is composed of chemically distinct subunits. A self-trapped carrier is then presumably confined to some subunit (e.g. a planar ring) of this multi-component molecule. Then maximal electronic overlap between corresponding structures of neighboring molecules requires their alignment with one another. As a result, intermolecular transport can be tortuous among a disordered distribution of such dopant molecules embedded within a medium.

Polarons also may form on large molecules comprising a symmetric array of nearly equivalent structures. For example, the linear pentacene molecule, illustrated in Fig. 6.3b, is based on the merger of five benzene molecules. In these situations charge may be distributed nearly uniformly over the chemically equivalent portions of the molecule. As described in Section 4.2, intramolecular movement of a weakly localized charge carrier in response to atomic motion is then prone to significantly shift the stiffness and frequencies of the associated atomic vibrations. As will be discussed in Chapters 11 and 12, carrier-induced changes in atomic frequencies can make significant contributions to a charge carrier's mobility and Seebeck coefficient, respectively. Indeed, the large Seebeck coefficients observed in pentacene transistors have been attributed to carrier-induced changes of the stiffness of associated atomic vibrations (von Mühlenen *et al.*, 2007).

7

Bipolarons

A *bipolaron* is a composite quasi-particle comprising (1) two charge carriers bound within a common self-trapping potential well (2) taken together with the displaced atomic equilibrium positions that produce it. The self-trapped carriers of a *large* bipolaron encompass several elementary structural units. By contrast, a *small* bipolaron's two self-trapped carriers collapse to a single elementary structural unit. When the two carriers share a common state the Pauli principle requires that their spins assume an anti-parallel alignment to form a singlet. Furthermore, as illustrated in Fig. 7.1, the merger of more than two carriers is opposed by the requirement that additional carriers occupy higher lying states. All told, coalesced carriers usually are presumed to pair as singlet bipolarons.

The electron–phonon interaction fosters the merger of self-trapped carriers. In particular, polaron models show the depth of a self-trapping potential well doubling when it is occupied by two carriers. Thus, the self-trapping contribution to each carrier's potential energy is lowered by their merger. However, the tendency for self-trapped carriers to coalesce is opposed by their mutual Coulomb repulsion. If and when self-trapped carriers amalgamate depends upon the resolution of these competing effects.

This chapter addresses the conditions governing the formation of large and small bipolarons. This discourse emphasizes the roles of the long- and short-range electron–phonon interactions. With just the Fröhlich long-range electron–phonon interaction two equivalent uncorrelated carriers will be bound provided that the static dielectric constant is much larger than the high-frequency optical dielectric constant, $\varepsilon_0 > 2\varepsilon_\infty$. However, these carriers' bipolaron is not stable with respect to separation into two separate polarons. Nonetheless, stabilization of this bipolaron can be achieved by reducing the two carriers' mutual Coulomb repulsion or enhancing their electron–phonon interaction. The carriers' mutual Coulomb repulsion can be reduced by electron-correlation effects that keep the paired carriers away from

Fig. 7.1 The occupation of non-degenerate self-trapped states is depicted for the merger of two singlet bipolarons into a quad-polaron.

one another. The electron–phonon interaction can be enhanced with the addition of a short-range component. However, if the short-range component is too strong, a large bipolaron will collapse into a small bipolaron.

Hund's rule states that the exchange energy between two carriers occupying degenerate states stabilizes their forming a triplet. However, if the carriers possess strong enough electron–phonon interactions with symmetry-breaking degeneracy-lifting atomic displacements a singlet will be stabilized instead. Firstly, coupling carriers' energies to symmetry-breaking atomic displacements reduces their stiffness constants. The lowering of the concomitant vibrations' free energy may be enough to stabilize the formation of a "softening" singlet bipolaron. Secondly, these electron–phonon interactions may be strong enough compared with the exchange energy to produce degeneracy-lifting shifts of atoms' equilibrium positions which stabilize a singlet bipolaron. Such a bipolaron may be termed a Jahn–Teller bipolaron.

7.1 Large-bipolaron formation

The scaling approach of Chapter 4 is used to address the formation and stability of a singlet bipolaron having d-dimensional electronic carriers (Emin and Hillery, 1989). To begin, consider the energy functional for the bipolaron's adiabatic ground state:

$$E_2(L) = \frac{2T_e}{L^2} - \frac{4V_L}{L} - \frac{4V_S}{L^d} + \frac{U}{L}$$

$$= \frac{2T_e}{L^2} - \frac{U}{L}\left(1 - \frac{2\varepsilon_\infty}{\varepsilon_0}\right) - \frac{4V_S}{L^d}. \tag{7.1}$$

This energy functional is obtained by extending the energy functional for a single self-trapped carrier, Eq. (4.30), to that for a pair of self-trapped carriers occupying a common state. In particular, having a pair of carriers (1) doubles their net kinetic energy, (2) quadruples their net polaronic energy since each carrier experiences double the polaronic energy lowering, and (3) adds the pair's mutual Coulomb repulsion. Here Eq. (4.32) has been utilized to express V_L, the long-range Fröhlich contribution to the polaron energy, in terms of U, the mutual Coulomb repulsion of two equivalent carriers immersed in a medium whose optical dielectric constant is ε_∞:

$$V_L \equiv \frac{U}{2}\left(1 - \frac{\varepsilon_\infty}{\varepsilon_0}\right). \tag{7.2}$$

Establishing a minimum of $E_2(L)$ at an intermediate value of L corresponds to forming a large bipolaron. By contrast, small-bipolaron formation corresponds to the minimum of $E_2(L)$ collapsing to the smallest permitted value of L, L_c. Then the self-trapped carriers are confined to a single structural unit.

The electron–phonon interaction will be sufficient to bind two carriers into a bipolaron if the energy functional $E_2(L)$ becomes negative. Requiring that $E_2(L)$ of Eq. (7.1) be negative yields the condition for bipolaron formation:

$$\frac{2T_e}{L} < U\left(1 - \frac{2\varepsilon_\infty}{\varepsilon_0}\right) + \frac{4V_S}{L^{d-1}}. \tag{7.3}$$

In the absence of the short-range portion of the electron–phonon interaction ($V_S = 0$) binding can only occur if $\varepsilon_0 > 2\varepsilon_\infty$. This condition is not satisfied in covalent or common ionic materials but is fulfilled in materials whose especially displaceable ions make $\varepsilon_0/\varepsilon_\infty$ exceptionally large (e.g. in relaxor or conventional ferroelectrics).

As illustrated in Fig. 7.2, adding a progressively stronger short-range component of the electron–phonon interaction increases the binding of a $d = 3$ large bipolaron while decreasing its spatial extent. Moreover, when V_S becomes large enough the minimum of $E_2(L)$ collapses to L_c, indicating small-bipolaron formation.

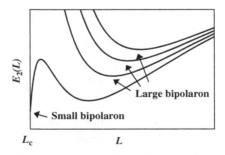

Fig. 7.2. The minimum of the bipolaron energy functional $E_2(L)$ at $L > L_c$, the smallest permitted value of L, corresponds to large-bipolaron formation. The large-bipolaron minimum deepens and its characteristic length decreases as the strength of the short-range component of the electron–phonon interaction V_S increases. The minimum of the energy functional ultimately shifts to L_c indicating the large bipolaron's collapse into a small bipolaron.

Beyond being bound, a realizable bipolaron should be stable with respect to separation into two independent polarons. To appreciate the physical content of this requirement, the bipolaron energy functional of Eq. (7.1) is rewritten in terms of that for two separated polarons (the square bracketed term):

$$E_2(L) = 2\left[\frac{T_e}{L^2} - \frac{V_L}{L} - \frac{V_S}{L^d}\right] + \frac{\varepsilon_\infty}{\varepsilon_0}\frac{U}{L} - \frac{2V_S}{L^d}. \tag{7.4}$$

Although the long-range electron–phonon interaction by itself produces a bound bipolaron if $\varepsilon_0/\varepsilon_\infty$ is sufficiently large, this $V_S = 0$ (Fröhlich) bipolaron remains unstable with respect to separation into two independent polarons (Pekar, 1963). In particular, the long-range electron–phonon interaction only reduces the net single-site repulsion between polarons from U to $(\varepsilon_\infty/\varepsilon_0)U$. This scaling-analysis result (Emin and Hillery, 1989) agrees with electrostatic considerations (Schultz, 1963).

It is evident from Eq. (7.4) that the addition of the short-range component of the electron–phonon interaction can stabilize a large bipolaron against its decomposition. However, the short-range electron–phonon interaction also fosters the collapse of a large bipolaron into a small bipolaron. The range of values of $2V_S/T_e$ for which a large bipolaron is stable against both decomposition and collapse widens as $\varepsilon_0/2\varepsilon_\infty$ increases beyond unity. For example, the range for which a planar ($d = 2$) large bipolaron is realizable is

$$\frac{4(\varepsilon_0/2\varepsilon_\infty) - 3}{2(\varepsilon_0/2\varepsilon_\infty)^2 - 1} < \frac{2V_S}{T_e} < 1. \tag{7.5}$$

The scaling approach has been used to study the energy of a pair of polarons as a function of their separation (Emin and Hillery, 1989). Even with stable bipolaron formation, the net energy of two polarons increases as their separation decreases until they overlap sufficiently to drive their merger into a bipolaron. In other words, there is an energy barrier to polarons coalescing into a bipolaron.

Bipolaron formation can involve correlating the positions of the two electronic carriers so that they can better avoid one another and thereby reduce their mutual Coulomb repulsion. However, keeping the two carriers away from one another also tends to reduce the polaronic binding that comes from the two carriers' merger.

Correlation effects are incorporated into a calculation of a bipolaron's energy by lifting the constraint that the bipolaron's two carriers occupy a common single-carrier orbital. For example, a correlated wavefunction, symmetric with respect to interchange of its two carriers, can be constructed with each carrier occupying a different single-carrier orbital. Correlated wavefunctions for one-dimensional large bipolarons were constructed by taking their two orbitals to have (1) different spatial extents and (2) their centroids displaced from one another (Emin, Ye and Beckel, 1992). A correlated wavefunction for a planar large bipolaron was constructed from two elliptical orbitals with their major axes perpendicular to one another (Emin, 1995a). The resulting four-lobed large bipolaron is stabilized when its electronic bandwidth is large enough.

Variational calculations, such as the two cited above, are used to address these refinements of the competition between carriers' mutual Coulomb repulsion and their mutual attraction induced by their electron–phonon interactions. The results of such computations are sensitive to choices of variational wavefunctions (Vinetskii and Giterman, 1958; Vinetskii, 1961; Suprun and Moizhes, 1982; Hiramoto and Toyozawa, 1985; Adamowskii, 1989; Verbist, Peters and Devreese, 1991; Basani, Geddo, Iadonisi and Ninno, 1991; Sil, Giri and Chatterjee, 1991). As such, qualitative features, rather than numerical results, appear most significant.

7.2 Small-bipolarons: negative-U centers

Charge carriers that collapse into small polarons readily localize, move by thermally assisted hopping, and typically manifest Curie paramagnetism. However, sometimes the expected Curie paramagnetism is not observed. One explanation of the absent Curie paramagnetism is that the carriers find it energetically favorable to localize as singlet pairs, small bipolarons, rather than as small polarons. Even then paired carriers tend to break apart as they hop to produce a relatively small thermally activated Curie paramagnetism (Emin, 1996).

Models of two carriers' collapse into a singlet small bipolaron often consider only the short-range electron–phonon interaction. Indeed, the long-range component of the electron–phonon interaction vanishes in non-polar materials where $\varepsilon_0 \to \varepsilon_\infty$. In this limit the energy functional for a singlet bipolaron, Eq. (7.1), becomes

$$E_2(L) = \frac{2T_e}{L^2} - \frac{4V_S}{L^d} + \frac{U}{L}. \tag{7.6}$$

The collapse of two carriers into a small bipolaron is described by this energy functional having its negative absolute minimum at the smallest acceptable value of L_c, the value of L corresponding to the two carriers merging at a single site. The condition for the collapse to a small bipolaron becomes: $(4V_S - U) > 4T_e/L_c$ for $d = 1$, $(4V_S - 2T_e)/L_c > U$ for $d = 2$, and $4V_S/(L_c)^3 > U/L_c + 2T_e/(L_c)^2$ for $d = 3$. All told, small-bipolaron formation requires (1) a strong-enough short-range electron–phonon interaction, (2) a sufficiently weak inter-carrier Coulomb repulsion, and (3) a sufficiently narrow electronic bandwidth. In the narrow-band limit, $T_e \to 0$, the requirement that a small bipolaron be bound can be rewritten simply as $4E_b > U_H$, where $E_b \equiv V_S/(L_c)^d$ is Holstein's small-polaron binding energy and $U_H \equiv U/L_c$ is the single-site Coulomb repulsion (Holstein, 1959b).

The difference between the small-bipolaron energy functional and that for two well-separated small polarons can be written in terms of E_b and U_H:

$$E_2(L_c) - 2\left(\frac{T_e}{L_c^2} - E_b\right) = -2E_b + U_H. \tag{7.7}$$

It is evident from Eq. (7.7) that this small bipolaron is energetically stable with respect to separating into two small polarons if $2E_b > U_H$. This short-range attraction between the two small polarons has been described as their having an attractive Hubbard interaction and as their having a negative-U (Pincus, 1972; Anderson, 1975).

Evidence for small bipolarons primarily consists of (1) observing carriers' hopping transport together with (2) the failure to observe the Curie paramagnetism expected of unpaired localized carriers. Such measurements generally occur in ionic and polar materials that are characterized by strong long-range electron–phonon interactions. In these materials the effective Coulomb repulsion between carriers is diminished by small ratios of optical-to-static dielectric constants. In particular, small-bipolaron hopping is said to occur in an intermediate-temperature ($\sim$130 K to $\sim$150 K) phase of single-crystal Ti_4O_7 (Lakkis *et al.*, 1976). These bipolarons are based on paired *d*-electrons of Ti^{3+} ions that are adjacent to one another. Hopping-type bipolarons are also reported in the low-temperature phase of slightly reduced crystalline WO_3 (Schirmer and Salje, 1980). Small bipolarons are also indicated in reduced- and doped-$BaTiO_3$ (Moizhes and Suprun, 1984; Kolodiazhnyi and Wimbush, 2006; Shao *et al.*, 2008).

Bipolaron formation is sometimes attributed to atoms of elements that manifest the "inert-pair effect". These atoms will not adopt valences that require breaking pairs of their outermost *s*-electrons. In particular, Tl can have valences of 1+ and 3+ but not 2+; Pb can have valences of 2+ and 4+ but not 3+; Bi can have valences of 3+ and 5+ but not 4+. For example, atoms of Bi in crystalline $BaBiO_3$ are idealized as alternating between Bi^{3+} and Bi^{5+} ions along the crystal's three principal directions. This structure is sometimes described as an ordered arrangement of Bi-based small bipolarons.

7.3 Softening bipolarons

At first glance boron carbides ($B_{12+x}C_{3-x}$ with $0.06 < x < 1.7$) appear as prime examples of materials whose carriers are small bipolarons. These materials' conductivities and Hall mobilities manifest thermally activated transport (Wood, 1986).

Moreover, the holes introduced by replacing carbon atoms with boron atoms do not produce Curie paramagnetism (Azevedo *et al.*, 1985). Furthermore, boron carbides also exhibit a small thermally activated paramagnetism attributable to breaking singlet pairs (Chauvet *et al.*, 1996). However, these materials do not display the customary polaron absorption, produced by exciting self-trapped carriers from the potential wells that bind them (Samara *et al.*, 1993). Moreover, the entropy transported with a charge carrier, measured by the Seebeck coefficient, is unusual as it is dominated by a very large contribution that is independent of the carrier concentration (Aselage *et al.*, 1998; 2001). The unusual absorption and Seebeck coefficient are suggestive of a "softening bipolaron".

Boron carbides are mixed crystals comprising $B_{11}C$ icosahedra and three-atom chains (Emin, 2006a). Electrons are donated from the chains to the electron-deficient icosahedra. Altering a chain's atomic constituents can change its donation to icosahedra thereby modifying the carrier density. Carriers move by thermally assisted hopping between icosahedra. The icosahedra are hollow spheroids whose frontier molecular orbitals have f-symmetry. An icosahedron's crystal field splits the seven-fold orbital degeneracy of these f-symmetry molecular orbitals into a low-energy four-fold-degenerate manifold and a higher energy three-fold-degenerate manifold. Electron donation in boron carbides ($B_{12+x}C_{3-x}$ with $0.06 < x < 1.7$) nearly fills the four-fold-degenerate manifold. In particular, the maximum hole concentration occurs near $x = 1$ yielding a singlet-bipolaron hole on about half of the icosahedra and a filled four-fold manifold on the remaining icosahedra.

The four-fold orbital degeneracy and two-fold spin degeneracy for each hole permits formation of many, $(4 \times 2)[(4 \times 2) - 1]/2 = 28$, different bipolaron states. Of these two-carrier states: four are same-state singlets, six are mixed-state singlets and 18 comprise triplets. Degeneracy among these two-carrier states is partially lifted by carriers' exchange interactions.

The concept of a softening bipolaron emerges upon investigating how a singlet bipolaron can be stabilized on an icosahedron (Emin, 2000b). In particular, as enunciated by Hund's rule, two carriers occupying degenerate orbitals are driven by their exchange interaction to align as a triplet rather than as a singlet. Nonetheless, as will become evident, static and dynamic symmetry-breaking deformations can overcome the exchange interaction and stabilize a singlet.

To appreciate how a singlet can be stabilized consider the algebraically simple example of two-fold orbital degeneracy (Emin, 2000b). The six two-carrier states of this example, $(2 \times 2)[(2 \times 2) - 1]/2 = 6$, consist of three singlets plus the three states of a triplet. Imposing a symmetry-breaking atomic distortion characterized by the parameter z can lower the electronic energy of the lowest-lying singlet W_S below the electronic energy of the triplet manifold W_T:

$$(W_S - W_T) = \left[3J - \sqrt{J^2 + (4F_z z)^2}\right]/2, \tag{7.8}$$

where J is the exchange splitting and F_z is the linear electron–phonon coupling constant associated with z. Without a deformation this singlet lies above the triplet: $W_S - W_T = J$ for $z = 0$. However, when the symmetry-breaking deformation is large enough the singlet state falls below the triplet, $W_S - W_T \to -2F_z z$ as $z \to \infty$.

That portion of the adiabatic potential which depends on symmetry-breaking is obtained upon adding its strain energy to its contribution to the lowest-lying singlet's electronic energy:

$$
\begin{aligned}
V_{ad,z}(z) &= \left[kz^2 + J - \sqrt{J^2 + (4F_z z)^2}\right]/2 \\
&= E_{bp,z}\left\{\left(\frac{kz}{2F_z}\right)^2 + \frac{2}{\gamma}\left[1 - \sqrt{1 + \gamma^2\left(\frac{kz}{2F_z}\right)^2}\right]\right\},
\end{aligned}
\tag{7.9}
$$

where $E_{bp,z} \equiv 4E_{p,z} \equiv 2F_z^2/k$ and $\gamma \equiv 4E_{bp,z}/J$. Here $E_{p,z}$ is the contribution to the localized polaron's binding energy from symmetry-breaking displacements of atoms' equilibrium positions. The symmetry-breaking contribution to the adiabatic potential $V_{ad,z}(z)$ is plotted in Fig. 7.3 for several values of γ. The stiffness of the minimum at $z = 0$, $k(1 - \gamma)$, is progressively reduced as γ is increased from zero to

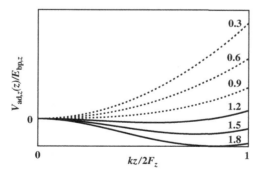

Fig. 7.3 The adiabatic potential for the lowest-lying singlet is plotted against the symmetry-breaking deformation parameter for six (indicated) values of the electron–phonon coupling parameter $\gamma \equiv 16E_{p,z}/J$. Here $E_{p,z}$ is the contribution to the localized polaron's binding energy from symmetry-breaking displacements of atoms' equilibrium positions and J is the singlet–triplet exchange splitting in the absence of distortion. The potential about the minimum at $z = 0$ softens as γ is increased toward unity. By contrast, the minimum shifts, deepens and stiffens as γ is increased above unity.

unity. As γ is increased beyond unity the minimum shifts to $z = (2F_z/k)(1 - 1/\gamma^2)^{1/2}$ as it concomitantly deepens to $-E_{bp,z}(1 - 1/\gamma)^2$ and stiffens to $k(1 - 1/\gamma^2)$.

These carrier-induced symmetry-breaking alterations of atomic vibrations change the coupled system's free energy by

$$F(\gamma, \omega) \equiv \kappa T \ln\left\{ \frac{\sinh[(\hbar\omega/2\kappa T)\sqrt{1-\gamma}]}{\sinh(\hbar\omega/2\kappa T)} \right\} \qquad (7.10)$$

for $\gamma < 1$ and by

$$F(\gamma, \omega) \equiv -\frac{J}{4}\frac{(\gamma-1)^2}{\gamma} + \kappa T \ln\left\{ \frac{\sinh\left[(\hbar\omega/2\kappa T)\sqrt{1-1/\gamma^2}\right]}{\sinh(\hbar\omega/2\kappa T)} \right\} \qquad (7.11)$$

for $\gamma > 1$. Figure 7.4 plots $F(\gamma, \omega)$ in units of $\hbar\omega$ against γ. It is evident that the free energy has minima at $\gamma = 1$ and falls with increasing γ when $\gamma \gg 1$. The free-energy minimum at $\gamma = 1$ is driven by lowering of the vibration frequency associated with symmetry-breaking atomic displacements. As such this softening minimum deepens with rising temperature. By contrast, the reduction of the free energy at $\gamma \gg 1$ results primarily from temperature-independent symmetry-breaking displacements of atoms' equilibrium positions. Higher electronic degeneracy and additional symmetry-breaking vibration modes enhance softening effects, broadening the softening region beyond that of this minimalist example (Emin, 2000b).

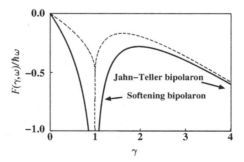

Fig. 7.4 $F(\gamma,\omega)/\hbar\omega$ is plotted against the electron–phonon coupling constant γ at two temperatures: $\kappa T = \hbar\omega/20$ (dashed curve) and $\kappa T = \hbar\omega$ (solid curve). The minimum of the free energy centered at $\gamma = 1$ primarily arises from lowering the frequency of the symmetry-breaking vibration. The bipolaron is dubbed a softening bipolaron when this lowering of vibrations' free energy is decisive in stabilizing it as a singlet. The fall of the free-energy at $\gamma \gg 1$ primarily arises from a symmetry-breaking shift of atoms' equilibrium positions. A Jahn–Teller bipolaron is stabilized primarily by symmetry-breaking shifts of atoms' equilibrium positions.

To stabilize a bipolaron formed among degenerate orbitals as a singlet rather than as a triplet requires that $-F(\gamma,\omega) > J$. The bipolaron is termed a softening bipolaron if its stabilization as a singlet results primarily from lowering atoms' vibration frequencies. By contrast, stabilization of a conventional or Jahn–Teller singlet bipolaron results mainly from shifting atoms' equilibrium positions.

The bipolaron will be energetically stable against separating into two polarons if

$$2E_{\mathrm{p}} + \left[-F(\gamma,\omega)-2E_{\mathrm{p},z}\right] > U, \tag{7.12}$$

where E_{p} and $E_{\mathrm{p},z}$ are the respective contributions to a polaron's binding energy from symmetry-conserving and symmetry-breaking displacements of atoms' equilibrium positions. In the absence of symmetry-breaking vibrations Eq. (7.12) simply reproduces the condition described below Eq. (7.7). The form of this condition is also reproduced in the strong-coupling limit $\gamma \gg 1$, where $-F(\gamma,\omega) \to E_{\mathrm{bp},z} = 4E_{\mathrm{p},z}$. In the complementary regime of Fig. 7.4 the reduced vibration frequencies of a singlet softening bipolaron, $\gamma = 16E_{\mathrm{p},z}/J \sim 1$ with $-F(\gamma,\omega) > J$, bolster the bipolaron's stability against decomposition into separate polarons since $-F(\gamma,\omega) \gg 2E_{\mathrm{p},z}$.

Distinctive features of boron carbides favor its charge carriers forming softening bipolarons (Emin, 2006a). In essence U is small (≈ 0.15 eV) because of electronic screening, $\varepsilon_\infty \approx 10$, and the bipolarons' two holes being confined to the surface of an icosahedron whose radius is large (~ 1.8 Å) and about 1/3 of carriers' mean separation. Conventional polaronic binding is also greatly reduced by carriers being spread over an icosahedron's 20 faces. Furthermore, softening effects are enhanced by the large size and four-fold orbital degeneracy of an icosahedron's frontier orbitals.

As will be discussed in Section 9.1 the optically assisted transfer of one of a small-bipolaron's electronic carriers to an unoccupied site produces an absorption band centered at $h\nu = 4E_{\mathrm{p}} - U$ for $4E_{\mathrm{p}} > U$ (Emin, 1993). This condition is always satisfied for a small bipolaron since its stability requires that $2E_{\mathrm{p}} > U$. However, such absorption need not occur for a softening bipolaron since its stability is primarily driven by lowering atoms' vibration frequencies, the square-bracketed term of Eq. (7.12). In particular, a stable softening bipolaron will not produce an absorption band if:

$$4E_{\mathrm{p}} < U < 2E_{\mathrm{p}} + \left[-F(\gamma,\omega)-2E_{\mathrm{p},z}\right], \tag{7.13}$$

where $-F(\gamma,\omega) \gg 2E_{\mathrm{p},z}$ for $\gamma \approx 1$. Thus, the absence of a polaron absorption peak in boron carbides is consistent with its carriers being softening bipolarons.

As detailed in Section 12.3 an electronic carrier's softening of atoms' vibrations produces a contribution to its Seebeck coefficient which is independent of the carrier concentration (Emin, 1999). Measurements of boron carbides' Seebeck coefficients find a very large contribution of this type (Aselage *et al.*, 1998; 2001). This distinctive finding is also consistent with boron carbides' carriers forming softening bipolarons.

8

Magnetic polarons and colossal magnetoresistance

Prior discussion addressed the effects of a carrier's energy depending on the positions of surrounding atoms. In a magnetic solid a carrier's energy also depends on exchange interactions that are governed by the relative alignments of a carrier's spin and unpaired spins of a host material's magnetic ions. Through these interactions a carrier influences the alignment of the host's unpaired spins. Meanwhile, the alignments of the host spins affect the carrier's state.

A *magnetic polaron* forms as a charge carrier added to an antiferromagnetic insulator (e.g. EuTe) self-traps inducing alignment of the associated spins into a ferromagnetic cluster. A magnetic polaron can be identified by its large and distinctive magnetic susceptibility. A high enough carrier density can even convert an antiferromagnet into a global ferromagnet. Here magnetic-polaron formation is addressed with a scaling analysis.

The scaling analysis is also applied to the free-energy arising from electron–phonon and exchange interactions of a donor electron in a ferromagnetic insulator (e.g. EuO). Exchange interactions between a large-radius donor electron and the localized spins of a ferromagnetic insulator oppose their thermally induced deviations from full alignment. At high enough temperature the cost in free energy of suppressing these spin deviations can become large enough to drive a donor to abruptly shrink to a bound small polaron with a concomitant loss of spin alignment. The temperature of this collapse rises in an applied magnetic field since it also suppresses spin deviations. Furthermore, direct electrical conduction between donors can be squelched by their collapse. In particular, with an appropriate density donors' thermally induced collapse can suppress their metallic impurity conduction to reveal the material's residual semiconducting behavior. The increase of the temperature of such a metal-to-semiconductor transition with an applied magnetic field is described as *colossal negative magnetoresistance* (CMR).

8.1 Magnetic polarons

A charge carrier usually self-traps by shifting the equilibrium positions of surrounding atoms. In principle a carrier could also self-trap by altering the ground-state alignment of the localized spins of a magnetic semiconductor (De Gennes, 1960). In particular, a carrier added to an antiferromagnetic insulator could realign associated spins to produce a ferromagnetic cluster. The proposed quasi-particle has been termed a "magnetic polaron" (Nagaev, 1967; Kasuya *et al.*, 1970; Umehara and Kasuya, 1972) and a "ferron" (Nagaev, 1976).

Magnetic-polaron formation is indirectly driven by the intra-site exchange interaction between a carrier and the magnetic ion it occupies. The direct effect of this interaction is to foster parallel or anti-parallel alignment of the carrier's spin with that of the occupied magnetic ion. As a result, misalignment of the local moments of different magnetic ions impedes carriers' movement between them (Anderson and Hasegawa, 1955). In particular, when the net intra-site exchange energy is much larger than the inter-site transfer energy the carrier's spin will align with that of the occupied ion. Then electronic transfer between two magnetic ions of spin S is reduced by a factor which falls from unity to $1/(2S + 1)$ as their misalignment is increased. In the classical limit ($S \rightarrow \infty$) this reduction factor corresponds to $\cos(\theta/2)$, where θ is the misalignment angle. In the complementary limit, when the net intra-site exchange energy is much less than the intersite transfer energy, the carrier maintains its spin alignment as it moves between magnetic ions. As a result, a carrier aligned with the magnetic moments of a ferromagnetic cluster encounters an energy barrier upon attempting to exit the ferromagnetic cluster. All told, intra-site exchange interactions tend to confine a carrier to a ferromagnetic cluster embedded within an antiferromagnetic insulator.

Antiferromagnets form with equal and opposite alignments of the spins of adjacent (1) ferromagnetic planes, (2) ferromagnetic chains or (3) individual magnetic ions. These ferromagnetic structures have dimensionalities of $d_f = 2$, 1, or 0, respectively. With magnetic-polaron formation a carrier converts some antiferromagnetically aligned spins to ferromagnetic alignment. A saturated ferromagnetic core might be surrounded by a halo of gradually diminishing unsaturated ferromagnetic alignments (Mauger and Mills, 1985).

For simplicity nonetheless presume a sharp demarcation between the region of saturated carrier-induced ferromagnetic alignment and the host material's antiferromagnetism as illustrated in Fig. 8.1. The ground-state energy functional of such a magnetic polaron in an electronically isotropic antiferromagnetic insulator with d_f-dimensional ferromagnetic constituents is modeled as the sum of three contributions (Emin and Hillery, 1988):

↓↑↓↑↓|↑↑↑↑↑↑↑↑↑↑↑↑|↓↑↓↑↓

Fig. 8.1 A magnetic polaron (in the rectangle) forms in an antiferromagnetic insulator as a carrier of charge e induces ferromagnetic alignment of the spins surrounding it.

$$E_{\text{mag},d_f}\left(L_a, L_f\right) = \frac{T_e}{3}\left(\frac{3-d_f}{L_a^2} + \frac{d_f}{L_f^2}\right) - \frac{|I|S}{2} + J_{\text{ex}}S^2\left(L_a^{3-d_f} L_f^{d_f}\right). \tag{8.1}$$

In Eq. (8.1) L_a and L_f are respectively the carrier's spatial extents (measured in units of the lattice constant) perpendicular to and parallel to an antiferromagnet's ferromagnetic planes, $d_f = 2$, or chains, $d_f = 1$. In addition, T_e is the carrier's half-bandwidth; I is the on-site exchange energy linking the spin-1/2 carrier with an occupied magnetic ion with spin magnitude S; and J_{ex} is the antiferromagnetic inter-ion exchange energy. The first contribution in Eq. (8.1) depicts the increase of a carrier's kinetic energy that results from containing it within the region of aligned spins. The second contribution is the net reduction of the energy of the carrier resulting from its intra-site exchange interactions with ferromagnetically aligned magnetic moments. This term is independent of the cluster's size because a carrier's interaction with each local moment is inversely proportional to the cluster size. The final contribution is the net increase of inter-ion exchange energy arising from flipping some of the antiferromagnet's spins to produce a ferromagnetic cluster.

The ground-state energy is found by minimizing the magnetic-polaron energy functional Eq. (8.1) with respect to L_a and L_f. At the minimum

$$L_a = L_f = L_{\text{mag}} \equiv \sqrt[5]{2T_e/3J_{\text{ex}}S^2} \tag{8.2}$$

with

$$E_{\text{mag}} = -\frac{|I|S}{2} + \frac{5}{3}\frac{T_e}{L_{\text{mag}}^2}. \tag{8.3}$$

The magnetic polaron's spatial extent L_{mag} is determined by the ratio of the carrier's confinement energy (characterized by T_e) to the inter-ion exchange energy $J_{\text{ex}}S^2$.

The magnetic polaron will be energetically stable if E_{mag} is less than the net energy in the absence of carrier-induced realignments of the antiferromagnetic's local moments. In the limit $T_e \gg |I|S/2$ the carrier maintains its spin as it moves between oppositely aligned sites. A carrier in the unaltered antiferromagnet is then delocalized with energy $E_{\text{free}} = 0$. The magnetic-polaron's stability condition becomes just $E_{\text{mag}} < 0$, where Eq. (8.3) and the condition $T_e \gg |I|S/2$ insures that the magnetic polaron be large: $L_{\text{mag}} > (10T_e/3\,|I|\,S)^{1/2} \gg 1$. Expressed alternatively, magnetic-polaron formation typically requires especially weak antiferromagnetic exchange:

$$T_e > \frac{|I|S}{2} > \frac{5}{3}\left[T_e^3\left(3J_{\text{ex}}S^2/2\right)^2\right]^{1/5}. \tag{8.4}$$

Applying a magnetic field tends to resize and reorient magnetic polarons' ferromagnetic clusters (Emin and Hillery, 1988). These effects yield huge magnetic susceptibilities for large ferromagnetic clusters. Thus a magnetic polaron should produce a very large magnetic susceptibility. Such an observation in doped EuTe is attributed to magnetic-polaron formation induced by wide s-band carriers' strong exchange with $S = 7/2$ f-band local moments within a very low-Néel-temperature antiferromagnet, $T_N \approx 9$ K (Vitins and Wachter, 1975). However, these observations may also indicate a ferromagnetic cluster formed by a carrier bound to a dopant, termed a bound magnetic polaron (BMP), rather than an intrinsic magnetic polaron (Mauger and Mills, 1985).

Electron–phonon interactions increasingly affect a magnetic polaron as its size is diminished (Umehara, 1981; Emin and Hillery, 1988). Indeed, like the situations described in Section 4.5, sufficient electronic localization of a magnetic-polaron's carrier can trigger its collapse into a small polaron. Similar effects come to the fore in the next section. Thermally produced spin deviations can destabilize a bound magnetic polaron in a ferromagnet with respect to its collapsing into a bound small polaron.

8.2 Donor-state collapse in ferromagnets: colossal magnetoresistance

Electron–phonon interactions can be pivotal in producing dramatic electronic transitions in magnetic semiconductors. In particular, as the local moments of a ferromagnetic semiconductor increasingly deviate from full ferromagnetic alignment a donor's electron ground state can abruptly collapse from being large to being severely localized. Such a sudden transition can occur because a short-range electron–phonon interaction only permits an extreme dichotomy in a carrier's localization. That is, a donor's electron can be so extended that the electron–phonon interaction has only a modest effect on its localization. Alternatively, the electron–phonon interaction can induce the donor's electron collapse to the smallest size compatible with a solid's atomicity and chemistry.

Consider a large-radius shallow donor added to a ferromagnetic insulator. Here exchange interactions between the magnetic insulator's local magnetic moments induce their ferromagnetic alignment. The exchange interactions of the donor electron with the ferromagnet's magnetic ions also foster their ferromagnetic alignment. The drive toward ferromagnetic alignment is therefore greatest in the vicinity of the donor electron. As a result, as depicted in Fig. 8.2, the donor electron tends to forestall thermally induced deviations of the local moments from full ferromagnetic alignment.

As discussed in Chapter 4, the short-range component of the electron–phonon interaction supports two qualitatively different ground states. In one instance the carrier extends over multiple sites. In the other situation the carrier collapses to the smallest size compatible with the solid's local chemistry. Direct extension of the scaling analysis of Section 4.3 yields the energy functional governing the ground state of a donor electron in a non-magnetic solid:

$$E_{donor}(L) = \frac{T_e}{L^2} - \frac{(V_L + V_{donor})}{L} - \frac{V_S}{L^3}. \tag{8.5}$$

Here the scaling constant produced by the donor electron's Coulomb attraction to its center is

$$V_{donor} \equiv C_{donor} \int dr |\phi_n(\mathbf{r})|^2 \left(\frac{1}{r}\right) \tag{8.6}$$

and C_{donor} governs the strength of this attraction. The donor collapses when the minimum of the energy functional occurs at the smallest acceptable value of

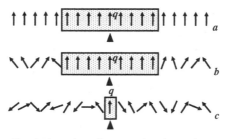

Fig. 8.2 The thermally induced collapse of a bound magnetic polaron in a ferromagnetic semiconductor is illustrated. (a) At very low temperatures the carrier (in the rectangle) bound to a defect (black triangle) does not perturb the alignment of the ferromagnet's spins. (b) At higher temperatures the carrier suppresses thermally induced deviations of surrounding spins from ferromagnetic alignment. (c) At high enough temperatures the carrier collapses thereby releasing surrounding spins to deviate from ferromagnetic alignment.

L, $L = L_c$. The donor's ground state is extended when the minimum of the energy functional occurs at a larger value of L.

A magnetic solid's donor electron also has intra-site exchange interactions with the local magnetic moments it contacts. Through these interactions the electron of this bound magnetic polaron (BMP) limits magnetic ions' deviations from full ferromagnetic alignment. In particular, the donor electron suppresses magnetic ions' thermal deviations from ferromagnetic alignment. As depicted in Fig. 8.2 the suppression is greatest for large-radius states. Thus, as the ferromagnet's temperature is raised toward its Curie temperature the reduction of the free energy from magnetic ions' deviations from ferromagnetic alignment increasingly fosters reducing the donor electron's spatial extent.

The free energy has been calculated with the ferromagnet's spin deviations being described as spin waves (Emin *et al.* 1986; 1987), valid below about 75% of the Curie temperature (Low, 1963). That portion of the calculated free energy which depends upon the spatial extent of the donor electron is then given by

$$F_{BMP}(L, T) = \frac{T_e}{L^2} - \frac{(V_L + V_d)}{L} - \frac{V_S}{L^3} - C\left(\frac{I}{\kappa T_C}\right)^2 \left(\frac{\kappa T}{L^2}\right), \qquad (8.7)$$

where T_C is the ferromagnet's Curie temperature and C is a numerical constant. The temperature-dependent contribution to $F_{BMP}(L, T)$ drives the BMP to shrink with rising temperature.

If the short-range component of the electron–phonon interaction is strong enough it produces a dichotomy between large-radius and collapse minima. Then, as illustrated in Fig. 8.3, the free-energy minimum can abruptly shift from that of a large-radius donor to that of a collapsed donor as the temperature is raised (Emin *et al.*, 1986; 1987). This thermally induced sudden collapse of a BMP is also depicted schematically in the lowest panel of Fig. 8.2. The difference between the entropy associated with spin deviations for large-radius and collapsed states underlies this thermally induced transition. The product of this entropy difference and the transition temperature is the latent heat accompanying the BMP's sudden collapse.

Applying a magnetic field suppresses spin deviations and thereby raises the temperature of the BMP collapse. In particular, the temperature of the collapse rises as the square root of the applied field (Hillery *et al.*, 1988). As a result, even a modest field can significantly raise the temperature of the collapse.

The collapse of BMP donors can dramatically affect direct transport between them. In particular, insulating behavior always prevails if the concentration of BMP donors is sufficiently small. Furthermore, metallic transport always results for large enough concentrations of BMP donors. At intermediate concentrations the

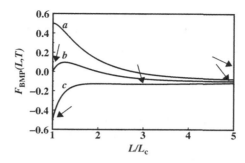

Fig. 8.3 The free-energy functional for a bound magnetic polaron in a ferromagnetic semiconductor, Eq. (8.7), is plotted against the ratio of the carrier's spatial extent L to that for a collapsed carrier L_c at three successively higher temperatures: curves a, b and c, respectively. Arrows point to the minima of these curves. The minimum at $L > L_c$ shrinks with rising temperature. Meanwhile the collapsed state's minimum $L = L_c$ emerges and is then stabilized.

thermally induced collapse of BMPs can shift their electrical transport from metallic to insulating. Applying a magnetic field raises the temperature of the collapse and the corresponding metal-to-insulator transition. At temperatures somewhat above the zero-field metal-to-insulator transition temperature application of the magnetic field thereby converts an insulator to metallic conduction. In other words, the rise of the metal-to-insulator transition temperature with magnetic field strength produces a huge negative magnetoresistance.

This discussion illustrates how electron–phonon interactions might contribute to abrupt transitions that occur in ferromagnetic semiconductors. In particular, this approach may be relevant to the dramatic metal-to-insulator transition observed in doped EuO.

Undoped EuO is a ferromagnetic semiconductor with a Curie temperature of 69 K (Methfessel and Mattis, 1968). It has the highest Curie temperature and smallest lattice constant of a series of structurally similar Eu-chalcogenide ferromagnetic crystals: EuO, EuS and EuSe. Ferromagnetism is thought to arise because direct ferromagnetic exchange between magnetic rare-earth ions dominates oxygen-mediated antiferromagnetic super-exchange. By contrast, antiferromagnetism dominates for EuTe which has the largest lattice constant.

N-type doping of EuO occurs through the introduction of oxygen vacancies (Oliver *et al.*, 1970; Torrence *et al.*, 1972; Godart *et al.*, 1980). Donors are also introduced by the substitution of Gd for Eu (Godart *et al.*, 1980). Samples that are too highly doped (> 1%) manifest metallic conduction without a metal-to-insulator transition (Godart *et al.*, 1980). Samples with intermediate doping levels show a dramatic transition from a low-temperature conducting state to a high-temperature insulating state at about 50 K, significantly below the Curie temperature (Oliver *et al.*, 1970; Penny *et al.*, 1972; Torrence *et al.*, 1972; Oliver *et al.*, 1972; Godart *et al.*, 1980). While Hall-effect measurements imply that the mobility falls significantly upon entering the insulating state, a huge reduction of the carrier density accounts for most of the dramatic (13–15 order-of-magnitude) reduction of the electrical conductivity which occurs at the sharpest metal-to-insulator transition (Oliver *et al.*, 1970; Petrich *et al.*, 1971; Godart *et al.*, 1980).

The thermally induced metal-to-insulator transitions observed in *n*-type EuO have been modeled as being thermally induced Mott-transitions in which carriers increasingly localize at donors (Torrence *et al.*, 1972; Leroux-Hugon, 1972; Mott, 1974; Kübler and Vigren, 1975; Mauger, 1983). These transitions also have been interpreted as being due to the suppression of metallic impurity conduction which accompanies donor states' collapse (Emin, 1986; Emin *et al.*, 1987; Hillery *et al.*, 1988). As illustrated in Fig. 8.2, thermally induced carrier localization in a ferromagnet lowers the free energy by enabling local magnetic moments to deviate from ferromagnetic alignment. As described in Fig. 8.3, the short-range component of the electron–phonon interaction enhances the severity of carriers' localization and thereby sharpens the transition.

Part II

Polaron Properties

9

Optical properties

9.1 Absorption from exciting polarons' self-trapped carriers

With polaron formation self-trapped electronic carriers are bound within potential wells established by shifts of atoms' equilibrium positions. Optical absorption attributed to polarons results from photo-excitation of their bound electronic carriers. The frequencies of these absorbed photons typically exceed atomic-vibration frequencies. As a result, one can invoke the Franck–Condon principle and presume that atoms remain fixed during the photo-excitation process (Franck, 1926; Condon, 1926). Thus a polaron's optical absorption simply probes photo-excitation of its self-trapped carrier to its excited electronic energy levels (Emin, 1993). The spectra of excited electronic states for large and small polarons differ qualitatively from one another to produce distinguishable optical absorptions.

A large polaron's self-trapped carrier extends over multiple structural units since its electronic bandwidth exceeds the depth of the potential well which binds it. In these circumstances the self-trapping medium can be treated as a deformable ionic continuum rather than as a solid composed of discrete ions. The eigenvalue spectrum for the self-trapped electron then resembles that for a donor electron in a conventional semiconductor.

As illustrated in Fig. 9.1, the electron can be photo-excited from the ground state of its Coulombic self-trapping potential well to unbound states associated with the conduction band. The threshold for this Franck–Condon absorption is just the electronic portion of the large polaron's binding energy. A large polaron's linear long-range electron–phonon interaction and its self-trapped carrier's Coulombic potential well produce simple numerical relations between contributions to its net binding energy E_p. In particular, net binding results from lowering the ground-state carrier's potential energy by $4E_p$ while increasing the material's strain energy by $2E_p$ and the carrier's confinement energy by half of the net lowering of the potential

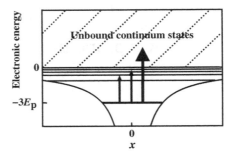

Fig. 9.1 Bound and unbound electronic energy levels of a large-polaron's self-trapped carrier are depicted as horizontal lines and a hatched area, respectively. Absorption processes for a large polaron include photo-excitation of its self-trapped carrier from its ground state (thick horizontal line) (1) to unbound continuum states (thick vertical arrow) and (2) to excited bound states (thinner vertical arrows).

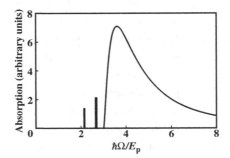

Fig. 9.2 A large polaron's absorption coefficient (in arbitrary units) is plotted against the energy of the exciting photon in units of the large-polaron binding energy, E_p. Photo-excitation of the self-trapped carrier to unbound continuum states produces a broad asymmetric absorption band. Photo-excitations of the self-trapped carrier to excited bound states produce relatively narrow absorption bands below the threshold energy $3E_p$.

energy $(4E_p - 2E_p)/2 = E_p$ (Emin, 1993). Therefore the threshold for photo-dissociation of the large polaron, the electronic portion of its binding energy, is $3E_p$.

A photon energy $\hbar\Omega$ above this threshold imparts kinetic energy of $\hbar\Omega - 3E_p$ to the freed electron (Emin, 1993). Moreover, the probability of exciting a self-trapped electron into the conduction band (1) increases with its density-of-states and (2) decreases rapidly when the wavevector of the freed electron exceeds $1/R_{LP}$, the reciprocal of the radius of the self-trapped electron. Thus, as illustrated in Fig. 9.2, the absorption coefficient arising from freeing a large polaron's self-trapped carrier is an asymmetric peaked function of photon energy whose width is typically much less than that of the conduction band.

Photons with energies below the threshold energy for liberating self-trapped carriers can also be absorbed. As illustrated in Fig. 9.1, photo-absorption can elevate a self-trapped carrier to an excited bound state of its self-trapping potential well. As depicted in Fig. 9.1, some of the potential photo-excitation processes may be forbidden by symmetry-based selection rules.

A small polaron's self-trapped carrier collapses to a single site with its binding energy exceeding its electronic bandwidth. In these circumstances, it is expedient to treat a small polaron as localized at one of a solid's discrete sites (Emin, 1993). As illustrated in Fig. 9.3, absorption of a photon can excite this self-trapped carrier from a low-lying state on one site to a higher energy state on a different site. In addition, a photon of equal energy can be emitted in the reverse process: transferring a carrier from the high-energy state on one site to a lower-lying state on a different site. The photon-absorption process dominates in thermal equilibrium since the carrier is then more likely to be found in the low-energy configuration than in the high-energy configuration.

For vanishing photon energy, $\Omega = 0$, the net absorption is simply proportional to the small-polaron's dc electrical conductivity (Klinger, 1963; Reik and Heese, 1967; Emin, 1975a). As the photon energy is increased the net absorption rises for the strong electron–phonon coupling characteristic of small-polaron formation. The absorption reaches its peak when the photon's energy equals the electronic energy needed to promote a small-polaron's self-trapped carrier from its ground state to an adjacent site while (in accord with the Franck–Condon principle) freezing atoms' positions.

The absorption peak's width is produced by atoms' departures from their ground-state equilibrium positions. As such, the absorption peak is most narrow at very low temperatures where atoms' zero-point motions predominate. As illustrated in Fig. 9.4 a small-polaron absorption band broadens with increasing temperature.

With small-bipolaron formation two carriers share a common site. Each carrier experiences an equilibrated self-trapping potential well that is twice as deep as that for an equilibrated small polaron's single carrier. Thus the net electronic energy of the Holstein small-bipolaron of Section 7.2 becomes $-8E_b + U$, where U is the two

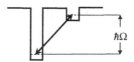

Fig. 9.3 A small-polaron's photo-absorption usually transfers its severely localized carrier from a state at one site to a higher lying state at a neighboring site. Photons also can be emitted whenever a localized carrier can be transferred from a high-energy state on one site to a low-lying state on an adjacent site.

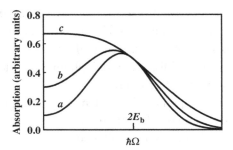

Fig. 9.4 A typical small-polaron absorption coefficient is plotted versus photon energy $\hbar\Omega$ for three temperatures. (*a*) At low temperature the absorption forms a band that peaks when $\hbar\Omega$ nears the difference between the electronic energy at an equilibrated unoccupied neighboring site and the electronic energy of the equilibrated self-trapped carrier. In the Holstein model (Holstein, 1959b) this energy is $2E_b$, defined below Eq. (7.6). (*b*) At higher temperatures increased displacements of atoms from their equilibrium positions broaden the absorption band. (*c*) At very high temperatures broadening obscures the peaked behavior.

carriers' mutual on-site Coulomb repulsion. After optically transferring one of the small-bipolaron's carriers to an equilibrated unoccupied site the system's electronic energy becomes $-4E_b$. The difference between the final and initial electronic energies is the energy of the exciting photon $\hbar\Omega = -4E_b - (-8E_b + U) = 4E_b - U$. Since formation of a conventional small bipolaron requires that $2E_b > U$, the energy of its characteristic photon, the energy of its absorption peak, is always greater than that for a small polaron, $2E_b$: $4E_b - U > 2E_b$ for $2E_b > U$. By contrast, no absorption is possible if $U > 4E_b$. This circumstance occurs if the bipolaron is a "softening" bipolaron, stabilized by carrier-induced reduction of vibration frequencies rather than by carrier-induced displacements of atoms' equilibrium positions (Emin, 2000b).

9.2 Low-frequency absorption from polaron motion

In addition to absorption that excites polarons' self-trapped carriers, polarons' slow motion enables low-frequency absorption. Large-polarons' coherent motion produces a Drude-like absorption that falls with increasing frequency. Small polarons' phonon-assisted hopping in disordered media produces an absorption that typically rises with increasing frequency.

A large polaron moves very slowly with a very large effective mass, Eq. (4.39). In particular, a large polaron's movement follows that of the atoms responsible for its carrier's self-trapping. The concomitant large-polaron bandwidth, Eq. (4.40), is less than the phonon energy characterizing vibrations of the atoms responsible for self-trapping. As a result, conservation of energy for a large polaron precludes the one-phonon emission and absorption processes which dominate the scattering of a conventional carrier. Rather a large polaron and phonons scatter via a two-phonon process (Schüttler and Holstein, 1986). This scattering occurs as long-wavelength ($> R_{LP}$) phonons experience the region of softened vibrations surrounding a large polaron. With sufficiently dispersive phonons a large polaron "reflects" the long-wavelength ($> R_{LP}$) phonons it encounters. The transport relaxation time τ for this process is given by

$$\frac{1}{\tau} \approx \frac{\kappa T}{m_{LP}\omega R_{LP}^2} \approx \omega \left(\frac{\kappa T}{E_p} \right), \tag{9.1}$$

when the thermal energy exceeds the energy of these phonons, $\kappa T > \hbar\omega$. As a result, the Drude electrical conductivity resulting from coherent motion of these large polarons,

$$\sigma(\Omega) = \frac{\sigma(0)}{1 + (\Omega\tau)^2}, \tag{9.2}$$

is distinguished from that of conventional carriers by its extremely long relaxation time τ. The very long large-polaron relaxation time indicates the difficulty of phonons' scattering these very massive quasi-particles.

Disorder often accompanies small-polaron and small-bipolaron formation. Disorder is intrinsic in semiconducting glasses and is produced extrinsically by imbedding dopants, impurities or defects within crystalline semiconductors. At low enough temperatures dc transport is limited by difficult thermally assisted hops that transport carriers between localized regions of relatively easy movement. By contrast, carriers that move with minimal thermal assistance within these regions of relatively easy transport dominate the low-temperature frequency-dependent electrical conductivity. The net polarization conductivity is the sum of contributions from these regions. The real part of the complex conductivity at frequency Ω produced by a carrier restricted to a region denoted by index i is:

$$\text{Re}[\sigma_i(\Omega)] \propto \Omega \left[\frac{2\Omega\tau_i}{1 + (\Omega\tau_i)^2} \right], \tag{9.3}$$

where τ_i is the relaxation time for the carrier's polarization current. The square-bracketed term on the right-hand side of Eq. (9.3) peaks when $\Omega\tau_i = 1$. Thus the observation of a low-frequency ($< 10^6$ Hz) polarization conductivity indicates the presence of relatively conducting regions with slow relaxation, large values of τ_i.

The classic case of low-frequency hopping conductivity is that of low-temperature electron hopping between shallow impurities in doped silicon. Specifically, low-frequency (10^2–10^5 Hz) conductivity is observed for the low-temperature (1–5 K) hopping of charge carriers between very low densities (10^{15}–10^{17} cm^{-3}) of shallow impurities in compensated crystalline silicon (Pollak and Geballe, 1961). The very slow relaxation of the ac conductivity is attributed to long ($\gg 100$ Å) hops between isolated pairs of shallow impurities. In particular, the small electronic overlap between the sites involved in very long hops gives them very small inter-site electronic transfer energies. With sufficiently small electronic transfer energies ($<\hbar\omega$) hops become "non-adiabatic", cf. Chapter 11. The concomitant hopping rates are very small since they are proportional to the square of their electronic transfer energies. These long hops dominate the low-frequency ac conductivity of lightly doped compensated silicon.

Low-frequency conductivity is also observed for low-temperature small-polaron and small-bipolaron hopping (Austin and Mott, 1969; Samara *et al.*, 1993). In these instances the hopping distances are comparable to a lattice constant (<5 Å). Indeed, the electronic transfer energies are often large enough ($> \hbar\omega$) to support adiabatic rather than non-adiabatic hopping. The magnitudes of the transfer energies are then not paramount in determining carriers' hopping rates. Rather, small-polarons' or small-bipolarons' self-trapped carriers move very slowly at low temperatures because their motion requires significant movement of the atoms associated with their self-trapping (Emin, 1992). By contrast, a hop of a carrier between shallow impurity-states requires only minimal atomic motion since atomic relaxation surrounding large-radius states is relatively small.

Disorder causes the relaxation times from different polarization centers to diverge from one another. Dispersion of the relaxation times for low-temperature hopping between shallow impurities results primarily from a spread of hopping distances. Dispersion of the rates for adiabatic small-polaron and small-bipolaron hopping results from a spread of energy differences between the involved sites. Low-temperature small-polaronic hops between nearly degenerate sites are especially sluggish due to the very low density of low-energy acoustic phonon states (Emin, 1992). With sufficient dispersion of the rates governing slow hops, the net conductivity typically increases sublinearly with increasing frequency at very low frequencies: $\text{Re}[\sigma(\Omega)] \propto \Omega^{1-s}$ (Pollak and Pike, 1972). This behavior is widely reported for hopping transport in disordered materials. However, these observations by themselves usually establish neither the hopping mechanism nor the source of relaxation rates' dispersion.

9.3 Carriers' photo-generation, recombination and luminescence

In conventional high-mobility semiconductors electrons and holes produced by the absorption of super-bandgap photons generally separate and thereby contribute to the dc conductivity. Photo-conduction ends as carriers become trapped. Ultimately carriers of opposing signs reunite via recombination centers, defects or impurities at which the oppositely signed carriers are sequentially captured.

A very different situation often prevails when both photo-generated carriers move slowly enough that they collapse into small polarons. Then only a small fraction of the electron–hole pairs generated from weakly absorbed photons separate from one another to contribute to photo-conduction. Rather the unseparated carriers ultimately merge with their nascent partners in a process termed geminate recombination. These well-localized carriers often recombine without the assistance of extrinsic traps, the recombination centers required in conventional high-mobility semiconductors.

The probability of separating a photo-induced slow moving electron and hole in a semiconductor is often treated as analogous to the thermal dissociation of oppositely charged ions in a weak electrolyte (Onsager, 1934; 1938). The probability of separating the electron and hole is then expressed as $\exp(-R_c/s_0)$. Here $R_c \equiv e^2/\varepsilon_0\kappa T$ defines the carriers' Coulomb capture radius and s_0 denotes the characteristic separation of thermalized photo-generated carriers, a strongly increasing function of photo-excitation energy. With $R_c \gg s_0$ most of these photo-generated electron–hole pairs geminately recombine. Applying an electric field assists a pair's dissociation by driving its electron and hole in opposing directions. The carrier generation efficiency in an applied electric field of strength F is then written as $\exp(-R_c/s_0)[1 + qR_cF/2\kappa T]$. The parameters R_c and s_0 can be obtained from measurements of the slope and the intercept at $F = 0$ of the carrier-generation efficiency plotted versus the applied electric field. Small radiation-induced carrier-generation efficiencies and geminate recombination are reported in diverse unconventional semiconducting materials. Some examples are: anthracene, PVK (poly-N-vinylcarbazole) and amorphous selenium (Kepler and Coppage, 1966; Batt, Braun and Hornig, 1968; Hughes, 1971a; 1971b; Melz, 1972; Pai and Enck, 1975; Chance and Braun, 1976).

With sufficiently slow motion these separated pairs of carriers will self-trap to form oppositely charged small polarons. Oppositely charged small polarons interact directly through their Coulomb attraction and indirectly through interference of their atomic-displacement patterns. Interference occurs as both carriers strive to displace each of the material's atoms. Such interference causes the net energy of the two oppositely charged small polarons to depend on their separation. Destructive interference causes the net energy of the two oppositely charged small polarons to rise as their separation is reduced. The sum of this repulsion and the long-range

Coulomb attraction of two oppositely charged polarons is plotted against their separation in Fig. 9.5. This illustration shows the total energy having its minimum at a finite separation.

Photo-generated species also include excitations in which electron and hole do not separate from one another. A mutually bound pair of oppositely signed carriers that share a common spatial region is regarded as an *exciton*. A Wannier exciton has at least one of its two oppositely charged carriers extending among multiple sites. A Frenkel exciton has both carriers sharing a common structural unit (e.g. a molecule or a bond). The binding of both types of exciton is enhanced by the relaxation of surrounding atoms. Since an exciton has no net charge, it has minimal Coulomb interactions with distant ions. Thus, atomic relaxation about an exciton is primarily governed by the electron–phonon interaction's short-range component rather than its long-range (Coulomb) component. An exciton together with the atomic relaxation that accompanies it is termed a *self-trapped exciton*. Figure 9.5 indicates the energy of an exciton and that of the self-trapped exciton which forms after atoms relax about it.

Luminescence at progressively lower energies results as the electrons and holes of excitons, self-trapped excitons and oppositely charged small polarons recombine with the emission of photons. The recombination of an exciton's electron and hole tends to be relatively rapid because their overlap is relatively strong. Lower-energy luminescence from radiative recombination of self-trapped excitons tends to be postponed since their formation is delayed by the necessity of transcending their energy barriers to self-trapping described in Section 4.3. Still lower-energy

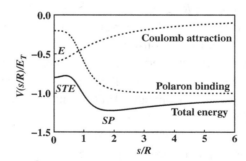

Fig. 9.5 The energy of two oppositely charged small polarons $V(s/R)$ is plotted (solid line) in units of their net binding energy E_T against their separation s divided by a small-polaron radius R. This energy sums the two polarons' Coulomb attraction with their polaron binding (dashed curves). Repulsion occurs as oppositely charged small polarons compete to displace surrounding atoms. The energy minima correspond to a self-trapped exciton (*STE*) and to separated small-polarons (*SP*). The minimum of the Coulomb attraction at $s/R = 0$ corresponds to the formation of an exciton (*E*).

luminescence from direct recombination of an electron-and-hole small-polaron pair tends to be slowest since their overlap with one another is relatively weak.

The luminescence spectrum from direct recombination of electron-and-hole small polarons evolves as their phonon-assisted hopping alters their separations. In particular, small-polarons' hopping facilitates their converging on their minimum-energy separations. In addition, as discussed in Section 9.1 and depicted in Fig. 9.3, irradiation within small-polarons' absorption bands induces small polarons' hopping. Indeed, small-polaron optical absorption bands are bleached out when absorption within these bands facilitates small-polarons' elimination via their recombination

Recombination occurs via phonon emission as well as via photon emission. These non-radiative and radiative recombination processes are illustrated using the generic configuration diagram of Fig. 9.6 (Dexter *et al.*, 1955). The upper curve depicts the energy of a deformable medium containing an electronic excitation plotted against x, a deformation parameter upon which it depends. In particular, the electronic excitation can be an exciton or separated electron-and-hole small polarons. The lower curve depicts the ground-state energy's harmonic dependence on this deformation parameter. The upward arrow at $x = 0$ represents photo-absorption of energy Δ_e by the undeformed ground state creating an electronic excitation. The subsequent displacements of atoms surrounding the electronic excitation correspond to the energy of the upper curve relaxing to its minimum at $x = x_m$. Luminescence of energy Δ_l from the excited-state minimum is represented by the downward arrow at $x = x_m$. The transition from the upper curve to the lower curve via their crossing point at energy E_{nr} and $x = x_{nr}$ corresponds to non-radiative recombination for classical atomic motion. With (1) a short-range linear electron–phonon interaction characterized by the force F and (2) both curves varying harmonically with stiffness constant k: $x_m = F/k$, $E_m = \Delta_e - F^2/2k$, $\Delta_l = \Delta_e - F^2/k$, $x_{nr} = \Delta_e/F$ and $E_{nr} = \Delta_e^2/(2F^2/k)$.

Recombination is a competition between radiative and non-radiative processes. Non-radiative recombination generally increases with electron–phonon interactions' strength. Luminescence results when electron–phonon interactions are ineffective in producing non-radiative recombination. This framework is adopted to explain why luminescence is observed from some F-centers but is not detected from others (Bartram and Stoneham, 1975).

The difference between the characteristic absorption energy and the luminescence energy is termed the luminescence Stokes shift. This energy generally heats a solid. In some instances (e.g. in alkali halides) this energy is imparted to atoms in such a way as to also efficiently produce structural defects (Pooley, 1966a; 1966b).

Irradiation that cools a material can also be envisioned (Pringsheim, 1929; Landau, 1946). As illustrated in Fig. 9.6, emission of a photon of energy Δ_l after

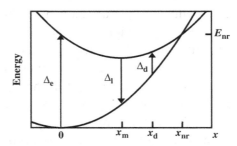

Fig. 9.6 Sums of the strain energy and the electronic energy for the electronic ground state (lower curve) and for an excited state (upper curve) are plotted against the deformation parameter x. The photo-excitation energy from the undeformed ground state is Δ_e. The energy of luminescence from the excited-state minimum at $x = x_m$ is Δ_l. The photo-excitation energy when the ground state is deformed to $x = x_d$ is Δ_d. The crossing point of the two curves at $x = x_{nr}$ establishes E_{nr}, the characteristic energy for a semiclassical non-radiative transition between the two states.

creation of an excitation by photo-exciting a deformed ground state with a photon of energy Δ_d extracts energy $\Delta_l - \Delta_d$ from the material. However, realizable optical cooling schemes require electron–phonon interactions that are sufficiently ineffective so that heating from ancillary Stokes shifts and non-radiative recombination are exceptionally small (Epstein *et al.*, 1995; Emin, 2007).

Optical properties of amorphous chalcogenide films appear consistent with the expectations of materials whose carriers form small polarons (Emin, 1980a). In particular, the observation that super-bandgap irradiation produces significant super-bandgap luminescence implies that relaxation of these electronic excitations is impeded by rapid localization (Shah and Bösch, 1979). Indeed, optically induced carriers manifest germinate recombination (Pai and Enck, 1975). Furthermore, time-resolved measurements find that optically induced carriers move with extremely low mobilities and recombine very slowly (Fork *et al.*, 1979). In addition, photo-conductivity measurements indicate an electron mobility that is too small to measure (Ing *et al.*, 1971). These results are consistent with steady-state charge transport being dominated by hole small polarons (Emin, Seager, and Quinn, 1972; Seager and Quinn, 1975). The nature of these carriers is suggested by the observation that super-bandgap excitations produce electron paramagnetic resonance (EPR) signals attributed to electrons on arsenic atoms and holes on chalcogen atoms (Bishop *et al.*, 1977). Super-bandgap irradiation also generates a mid-gap absorption band which together with the induced EPR signals can be bleached out with mid-gap irradiation (Bishop *et al.*, 1977). Recombination is primarily non-radiative with steady-state super-bandgap illumination producing a large luminescence Stokes shift of about half the semiconducting energy gap (Fischer *et al.*, 1971). Time-dependent luminescence measured after pulsed illumination reveals several bands that evolve to leave just the steady-state result at long times (Bösch and Shah, 1979). The time dependences of the three observed luminescence bands has been described in terms of exciton, self-trapped exciton and small-polaron radiative recombination (Bösch *et al.*, 1980). An alternative model for the electrical transport and optical properties of chalcogenide glasses proposes that they result from polaronic electron and hole traps associated with high densities of bonding defects (Street and Mott, 1975; Kastner *et al.*, 1976).

These findings are similar to those from transition-metal compounds in which charge carriers are thought to form small polarons. In particular, steady-state transport measurements on $LiNbO_3$ indicate that electrons form small polarons (Nagels, 1980). Moreover, super-bandgap irradiation produces a mid-gap absorption band and two EPR signals from two distinct types of localized spins, O^- and Nb^{3+} ions. (Schirmer and von der Linde, 1978; Schirmer *et al.*, 2009).

10

Large-polaron transport

10.1 Coherent versus incoherent transport

Electronic transport is often described in terms of two complementary models. Coherent transport envisions a carrier moving through a material with its motion occasionally disrupted by scattering events. By contrast, incoherent transport views a carrier as only making occasional jumps between localized regions. Coherent transport in conventional semiconductors is generally ascribed to a carrier with a mobility that greatly exceeds 1 cm^2/V-sec and falls with increasing temperature. Incoherent transport is often attributed to a carrier with a mobility that is much less than 1 cm^2/V-sec and rises with increasing temperature.

Treatments of coherent transport usually begin with a charge carrier in a delocalized state whereas incoherent transport starts with a carrier in a localized state. As a result, delocalized states are often said to produce coherent transport while localized states produce hopping transport. However, there is not a one-to-one correspondence between the nature of a carrier's eigenstates and its transport. The character of a carrier's transport also depends on its scattering. For example, periodicity ensures that the eigenstates of a small polaron in a crystal are Bloch-like delocalized states. Nonetheless, as will be discussed in Chapter 11, small-polaron transport typically proceeds by phonon-assisted hopping. Coherent transport is then destroyed as scattering reduces a small-polaron's mean-free-path ℓ to below its diameter $d_c < a$, the lattice constant: $\ell < d_c$. Thus, transport among delocalized eigenstates can be incoherent. In addition, transport among localized states can be coherent. For example, consider the Wannier–Stark localized electronic eigenstates formed when an electric field F is applied across a periodic potential of lattice constant a (Wannier, 1959; Emin and Hart, 1987a). The eigenvalues comprise a Wannier–Stark ladder whose successive rungs are separated in energy by eFa. The corresponding electronic eigenstates are each localized with the Wannier–Stark length of $L_{W-S} \sim W/eF$ about successive lattice sites, where W is the electronic bandwidth in the absence of the applied electric field. With $L_{W\text{-}S} \gg a$, many Wannier–Stark eigenstates strongly overlap one another even as their spatial oscillations ensure their mutual orthogonality. However, scattering mixes these states. Coherent transport results when the Wannier–Stark length exceeds the carrier mean-free-path, $L_{W\text{-}S} > \ell$ (Kane, 1959; Emin and Hart, 1987b). Thus, conventional semiconductors display coherent transport as evidence of localization is obliterated. By contrast, Wannier–Stark localization is indicated when $L_{W\text{-}S} \ll \ell$. This exceptional circumstance arises when a sufficiently strong electric field is applied to a narrow-band super-lattice whose carriers have a long mean-free-path (Mendez *et al.*, 1988).

In summary, coherent scattering-type transport can be envisioned when a carrier's mean-free-path exceeds its diameter d_c. Coherent transport even occurs among localized states provided that the carrier's mean-free-path is sufficient to disrupt the localization: $L_{loc} > \ell > d_c$. Incoherent transport occurs when scattering destroys the carrier's integrity: $\ell < d_c$. The forthcoming discussion will develop these notions further.

The Boltzmann equation is frequently used to derive an expression for a carrier's mobility μ in terms of its velocity v and mean-free-path ℓ. This relationship can be used to establish the minimum mobility for coherent transport. Specifically, this lower limit is found by replacing a carrier's mean-free-path by its characteristic length, d_c. It is then noted that a coherently moving carrier's diameter is related to its de Broglie wavelength λ: $d_c \equiv \lambda/2\pi = \hbar/p = \hbar/m^*v$, where p and m^* are the carrier's momentum and mass, respectively. The carrier mobility μ for scattering-type coherent transport is constrained by:

$$\mu = \frac{e}{\kappa T}v\ell > \frac{e}{\kappa T}vd_c = \frac{e}{\kappa T}\frac{\hbar}{m^*}. \tag{10.1}$$

Thus, for a conventional wide-band semiconductor, self-consistency of the standard Boltzmann-equation-based scattering treatment requires a high mobility (Fröhlich and Sewell, 1959; Herring, 1961). For example, the requirement of Eq. (10.1) evaluated at 300 K with m^* even as large as the free-electron mass is that $\mu > 45$ cm^2/V-sec. However, the Boltzmann-scattering approach can be applied to a much lower mobility if the carrier's effective mass is enhanced significantly above the free-electron mass. Such large effective masses occur with large-polaron formation. Room-temperature Hall mobilities of ~1–10 cm^2/V-sec that are reported for n-type carriers in alkali halides, alkaline-earth fluorides and some transition-metal oxides have been attributed to large polarons (Seager and Emin, 1970; Seager, 1971; Forro *et al.*, 1994).

10.2 Large-polaron effective mass

A self-trapped carrier can only move when the atoms responsible for its binding alter their positions. This movement of atoms generally changes the self-trapped carrier's potential well and therefore its energy. If the variation of the self-trapped carrier's energy during its motion remains less than its electronic band half-width $W/2$ (the energetic measure of its coherence) its motion is coherent.

The motion of a large-polaron's self-trapped carrier generally retains its coherence. In particular, in accord with Section 4.3, the energy of the self-trapped carrier of a large polaron is $E_{st} = -V_{def}^2/W$, where V_{def} is the deformation-related energy and the condition $W/2 > V_{def}$ insures that the polaron is large (Holstein, 1959a). The change of the self-trapped carrier's energy associated with its motion is $\Delta E_{st} = 2V_{def}\Delta V_{def}/W$, where ΔV_{def} is the change of its deformation-related energy associated with moving the carrier a lattice constant. Large-polaron motion will be coherent if $1 > |\Delta E_{st}|/(W/2) = V_{def}\Delta_{def}/(W/2)^2$. This condition is fulfilled since large-polaron formation and motion is generally conditioned by $W/2 > V_{def} \gg \Delta V_{def}$.

A large polaron's effective mass is often defined as the net shift of atomic mass associated with translating an equilibrated self-trapped carrier by a lattice constant. In particular, components of the effective-mass tensor $M_{j,j'}$ link the kinetic energy of a large polaron with components of its velocity in directions j and j'. That is, the large-polaron kinetic energy is

$$\text{K.E.} \equiv \frac{1}{2} \sum_{j=1}^{3} \sum_{j'=1}^{3} M_{j,j'} \left(\frac{\delta x_j}{\delta t} \right) \left(\frac{\delta x_{j'}}{\delta t} \right), \tag{10.2}$$

where δx_j is the distance that the self-trapped carrier is translated in direction j during time interval δt. Here the effective-mass tensor is defined by

$$M_{j,j'} \equiv \sum_{g} \sum_{i=1}^{3} M_g \left(\frac{\delta u_{g,i}}{\delta x_j} \right) \left(\frac{\delta u_{g,i}}{\delta x_{j'}} \right), \tag{10.3}$$

where $\delta u_{g,i}$ is the shift in direction i of the atom of mass M_g located at site g that moves the carrier δx_j in direction j and $\delta x_{j'}$ in direction j'. The shifts of atoms' equilibrium displacements after translating the carrier by a lattice constant in the j-direction equals the differences between equilibrium displacements in the negative j-direction of adjacent pairs of atoms: $\delta u_{g,i} \rightarrow \delta U_{g,i}$. Thus, the large-polaron's effective-mass tensor can be rewritten in terms of its static properties as:

$$M_{j,j'} = \sum_{g} \sum_{i=1}^{3} M_g \left(\frac{\delta U_{g,i}}{\delta a_j} \right) \left(\frac{\delta U_{g,i}}{\delta a_{j'}} \right), \tag{10.4}$$

where the two bracketed terms are elements of the polaron's carrier-induced strain tensor with δa_j denoting the lattice constant in the j-direction.

Equation (10.4) has been evaluated for different models of the electron–phonon interaction (Emin and Hillery, 1989). For the three-dimensional large-polaron formed with the Fröhlich long-range interaction the effective mass is [cf. Eq. (4.39)]:

$$m_{LP} \approx \frac{E_{LP}}{\omega^2 R_{LP}^2}, \tag{10.5}$$

where E_{LP} and R_{LP} are the large-polaron binding energy and radius, respectively, and ω is the optical-vibration frequency. The three-dimensional result has also been expressed in terms of the Fröhlich coupling constant α as $m\alpha^4$ with $\alpha \gg 1$, where m is the free-carrier mass and α is defined in Eq. (3.3) (Fröhlich, 1963). For a one-dimensional large polaron formed by a carrier's interaction with acoustic phonons the effective mass is $\sim 4E_{LP}/c_s^2$, where c_s is the sound velocity (Schüttler and Holstein, 1986). These formulae indicate that large-polaron effective masses are typically orders of magnitude greater than the free-electron mass.

A large polaron moves very slowly with a very large mass because its self-trapped carrier only moves when the atoms associated with its self-trapping alter their positions. Concomitantly, the large mass of a large polaron can only be observed by measurements performed at frequencies Ω well below that of vibrating atoms: $\Omega \ll \omega$ (e.g. at microwave frequencies). Higher-frequency measurements observe just the mass of the self-trapped electron not the mass of the large polaron. That is, such measurements primarily observe electronic motion within the self-trapped well.

The preceding discussion addressed the adiabatic effective mass of a large polaron associated with atoms' continuous classical motion. However, a second type of large-polaron motion is possible. The self-trapped carrier can follow atoms as they tunnel between discrete geometrically equivalent atomic-deformation patterns (cf. Fig. 10.1). By itself, such tunneling motion is coherent since it involves no energetic variations. However, atomic tunneling significantly reduces the coherence

Fig. 10.1 Displacements of atoms' equilibrium positions are schematically illustrated for a one-dimensional large polaron before (*a*) and after (*b*) it is translated by one lattice constant to the right. The disparity between an atom's displacement before and after translation is the same as that between the atom and its nearest-neighbor to its left.

energy (the half-bandwidth) for this coordinated electronic and atomic motion from that for pure electronic motion. Thus, this type of coherent transport will be squelched with sufficient energetic disorder and scattering.

The bandwidth for such large-polaron motion is proportional to the matrix element of the system's adiabatic Hamiltonian between atomic-vibration wave-functions associated with large polarons centered at different sites. Beyond having different centroids these two wavefunctions represent geometrically equivalent configurations of atomic equilibrium displacements and equal quantum numbers. The adiabatic Hamiltonian describes a self-trapped carrier among a periodic array of vibrating atoms. In accord with the adiabatic principle these atoms' equilibrium positions change in concert with the location of the self-trapped carrier. Thus, a large polaron's adiabatic potential comprises a periodic array of deformable potential wells. All told, the bandwidth associated with a large polaron tunneling between states with different centroids is a generalization of adiabatic small-polaron tunneling between the two sites of a double well (Holstein, 1959b).

Atoms' vibrations about their equilibrium positions are described by the product of displaced-oscillator wavefunctions. Translating atomic equilibrium positions of the displaced-oscillator wavefunctions coherently shifts a self-trapped carrier between geometrically equivalent positions. The overlap between the products of displaced-harmonic-oscillator wavefunctions that correspond to a self-trapped carrier shifted by the vector h is defined as $\exp[-S(h)]$, where the vibrations' quantum numbers have been replaced by their thermal average (Holstein, 1959b).

For definitiveness, consider large-polaron formation on a linear chain of displaceable atoms labeled by the index g that vibrate about the equilibrium displacement $u_{p,g}$ with the frequency ω and stiffness k when the self-trapped carrier is centered at site p (Holstein, 1959a). The overlap function for this model is

$$e^{-S(h)} = \exp\left[-\sum_{g=-\infty}^{\infty} \frac{k\left(u_{p,g}-u_{p,g-h}\right)^2}{4\hbar\omega} \coth(\hbar\omega/2\kappa T)\right], \qquad (10.6)$$

where the symmetry condition $u_{p+h,g} = u_{p,g-h}$ has been used. This overlap factor vanishes in the classical limit, $\hbar \to 0$, where atomic tunneling is prohibited. The overlap factor also falls with increasing temperature since the overlap between displaced harmonic-oscillator wavefunctions decreases with increasing vibration energy as additional nodes produce more cancellations. Finally, the overlap falls rapidly as h, the separation between the centroids of the overlapping large-polarons, is increased.

The nearest-neighbor transfer energy for this one-dimensional large-polaron is

$$t_{LP} \approx \sqrt{\frac{3E_{LP}(L)\hbar\omega}{\pi L}} \exp\left[-\frac{E_{LP}(L)}{\hbar\omega L^2}\coth(\hbar\omega/2\kappa T)\right], \qquad (10.7)$$

where L is the half-length of the large polaron expressed in units of the lattice constant and $E_{LP}(L) \propto 1/L$ is the large-polaron binding energy derived by Holstein (1959a). With pure nearest-neighbor transfer the one-dimensional large-polaron bandwidth from this tunneling motion is $W_{LP,tun} = 4t_{LP}$. The width of this large-polaron tunneling band is typically less than the phonon energy $W_{LP,tun} < \hbar\omega$ given the requirement for large-polaron formation $E_{LP}(L) \gg \hbar\omega$.

Distinctively, this tunneling bandwidth narrows with increasing temperature. This narrowing becomes most pronounced once κT rises above $\hbar\omega/2$. Then the large-polaron effective mass associated with tunneling motion, $\propto 1/W_{LP,tun}$, rises exponentially with increasing temperature. This feature has been suggested as a possible explanation of the steep fall of the large-polaron Hall mobility as the temperature is raised above the characteristic phonon temperatures of alkali halide and alkaline-earth fluoride crystals (Seager and Emin, 1970; Seager, 1971).

10.3 Large-polaron scattering

The dynamics of a large-polaron's motion are qualitatively different from those of an electronic charge carrier in a conventional semiconductor. For definiteness consider a one-dimensional large polaron which interacts with acoustic phonons (Schüttler and Holstein, 1986). The ratio of a large polaron's mass m_{LP} to that of a conventional electronic carrier m is huge: $\sim W_e E_{LP}/(\hbar\omega_D)^2 \gg 1$, where W_e is the conventional-carrier's electronic bandwidth, E_{LP} is the large polaron binding energy and ω_D is the Debye frequency. Furthermore, the thermal velocity of a large polaron is much less than that of a conventional carrier: $\sim (m/m_{LP})^{1/2}$. The large polaron's thermal velocity is even smaller than the sound velocity c_s: $\sim c_s(\kappa T/E_{LP})^{1/2}$ with $E_{LP} \gg \kappa T$. Moreover, the thermal momentum of a large polaron p_{th} is much larger than that of a conventional electronic carrier: $\sim (m_{LP}/m)^{1/2}$. The large polaron's thermal momentum even exceeds the thermal momentum of an acoustic phonon $\hbar k_{th} \equiv \kappa T/c_s$ by $(E_{LP}/\kappa T)^{1/2}$.

These dynamic considerations depict a large polaron moving slowly with a large momentum while scattering with phonons having relatively small momenta. By contrast, an electronic carrier in a conventional semiconductor usually moves rapidly with small momentum while scattering against phonons with relatively large momenta. For these reasons large polarons are only weakly scattered by phonons while conventional electronic carriers are strongly scattered by phonons.

A large polaron's scattering with phonons occurs as they encounter the region of softened vibrations that surrounds a self-trapped carrier. As depicted in Fig. 10.2, a self-trapped carrier adjusts to alterations of the positions of the atoms associated with its self-trapping. This electronic relaxation reduces the net energy expended in displacing the associated atoms thereby reducing their stiffness constants. Affected stiffness constants can be reduced by significant fractions [estimates may be made using Eqs. (4.15) and (4.18)]. The cross-section for a large polaron's scattering by phonons peaks at the resonance when the phonon wavevector is comparable to the reciprocal of the softened region's spatial extent (Schüttler and Holstein, 1986).

A large-polaron's scattering by phonons is unlike that of a conventional carrier. A large-polaron's huge effective mass precludes most processes in which a single phonon is emitted or absorbed since $m_{LP}c_s \gg p_{th} \gg \hbar k_{th}$. Thus a large-polaron's scattering is dominated by processes in which relatively little net energy is transferred between the phonons and the large polaron. By contrast, the width of the carrier's energy band in a conventional semiconductor is very much greater than phonon energies. Therefore scattering of a conventional carrier occurs primarily via the absorption or emission of single phonons.

The predominant process by which a large polaron is scattered by ambient phonons is illustrated in Fig. 10.3. A slow-moving large-momentum large polaron is scattered as it "reflects" faster-moving low-momentum phonons (Schüttler and Holstein, 1986). This process is analogous to that by which a macroscopic ball moving through air encounters resistance by colliding with relatively light fast-moving molecules.

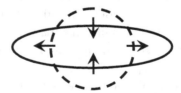

Fig. 10.2 A self-trapped carrier (dashed circle) alters its shape (to form an oval) in response to atoms' movements. The self-trapped carrier's polarization (emphasized with arrows) reduces the net stiffness governing these atomic displacements.

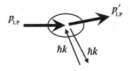

Fig. 10.3 The softened region of a large polaron (the ellipse) scatters incident phonons. The large momentum of a thermal heavy-massed large polaron enables it to reflect the relatively low-momentum phonons it scatters.

Consider how a one-dimensional large polaron of length L is scattered by ambient acoustic phonons indexed by wavevector k (Schüttler and Holstein, 1986). The large polaron's scattering rate is the thermal average of the product of (1) the density of ambient phonons $N(k/k_{th})$, (2) the cross-section for their being reflected by the large polaron $\sigma(kL)$, and (3) the resulting change of a large polaron's velocity $2\hbar k/m_{LP}$:

$$\frac{1}{\tau} = \left\langle N(k/k_{th})\sigma(kL)\left(\frac{2\hbar k}{m_{LP}}\right)\right\rangle_{thermal}, \tag{10.8}$$

where it is noted that a phonon transfers twice its momentum to a polaron from which it is reflected and $\hbar k_{th} \equiv \kappa T/c_s$ is the thermal-acoustic-phonon momentum. In the low-temperature limit the phonons' Bose factor $\approx \exp(-k/k_{th})$ restricts their scattering to those for which $kL \ll 1$ with $\sigma(kL) \propto (kL)^2$. The mobility then becomes

$$\mu_{LP} = \frac{e\tau}{m_{LP}} \propto \frac{eL^2}{\hbar}\left(\frac{T_L}{T}\right)^4, \tag{10.9}$$

where $\kappa T_L \equiv \hbar c_s/L \gg \kappa T$. In the regime $T \gg T_L$ the phonons' Bose factor $\approx k_{th}/k$ permits their scattering to be dominated by those for which $kL \approx 1$ with $\sigma(kL) \approx 1$ yielding

$$\mu_{LP} \approx \frac{eL^2}{\hbar}\left(\frac{T_L}{T}\right) = \frac{ec_s L}{\kappa T}. \tag{10.10}$$

Acoustic phonons' weak scattering of a one-dimensional large polaron, indicated by its long scattering time, confirm that its transport is coherent. In particular, even when $T > T_L$ the large-polaron's mean-free-path exceeds its spatial extent:

$$\ell = v_{LP}\tau \approx \left(\frac{2\kappa T}{m_{LP}}\right)^{1/2}\left(\frac{m_{LP}c_s L}{\kappa T}\right) = \left(\frac{8E_{LP}}{\kappa T}\right)^{1/2} L \gg L, \tag{10.11}$$

where it is noted that $\tau = m_{LP}c_s L/\kappa T$, cf. Eq. (10.10), and recalled that $m_{LP} \approx 4E_{LP}/c_s^2$.

The preceding discussion of this section has the large-polaron's very long scattering time compensating for its very big effective mass ($\tau \propto m_{LP}$) to produce a mobility that is independent of its effective mass. In fact, the magnitude of this large-polaron's mobility is comparable to that of conventional charge carriers: > 1–10 cm^2/V-sec at 300 K. In essence, the weak phonon scattering of this slow-moving massive large polaron gives a similar mobility to that from strong phonon scattering of a fast-moving conventional electronic carrier.

Distinctive aspects of this large-polaron's transport may nonetheless be observed. The unconventionally long large-polaron scattering time shifts the Drude tail of the frequency-dependent conductivity $\text{Re}[\sigma(\Omega)] \propto 1/[1 + (\Omega\tau)^2]$ to exceptionally low frequencies, toward the microwave regime [cf. Section 9.2]. Furthermore, the scattering of acoustic phonons by large polarons is unlike phonons' absorption and emission by conventional carriers. This distinctive feature may be discernable from studying the effect of carriers on phonons' thermal diffusivity.

Large-polaron transport is dramatically different for scattering with dispersion-less optic phonons (Holstein, 1981; Schüttler and Holstein, 1986). In the absence of optic-phonons' dispersion their group velocity vanishes. Then optic phonons behave as thermally generated stationary imperfections which scatter the large polaron. The mobility is inversely proportional to the density of these scattering centers. Thus the mobility garners a factor which falls with increasing temperature as $\exp(\hbar\omega/\kappa T)$ when $\kappa T \ll \hbar\omega$, the optic-phonon energy. Moreover, a large-polaron's scattering cross-section with dispersion-less optical phonons falls once its momentum exceeds $\hbar/L$ (Holstein, 1981; Schüttler and Holstein, 1986). This effect contributes a factor to the mobility that rises with increasing temperature as $\exp(\pi^2\kappa T/E_c)$ for $\kappa T \gg E_c$, the kinetic energy at which a large polaron's de Broglie wavelength $\hbar/p_{LP}$ equals its length L: $(p_{LP})^2/2m_{LP} = (\hbar/L)^2/2m_{LP} \equiv E_c \ll \hbar\omega$ [cf. Eq. (4.40)]. Thus, the mobility's temperature dependence results from a competition between (1) the number of scattering centers that increases with rising temperature and (2) the efficacy of these centers' scattering that falls with increasing temperature. The first factor dominates at lower temperatures to produce a mobility that falls with increasing temperature until $\kappa T = (E_c\hbar\omega)^{1/2}/\pi$. As the temperature is increased further the second effect begins to dominate causing the mobility to rise from its minimum value.

All told, a cogent view of a large polaron's scattering with phonons has emerged. Polarization of a large-polaron's self-trapped carrier in response to atoms' movements softens their vibrations. The softened region then scatters phonons. The nature of the resulting scattering depends critically on the large-polaron's velocity $(2\kappa T/m_{LP})^{1/2}$ relative to the phonons' group velocity $\partial\omega(k)/\partial k$. In one limit the relatively slow moving large polaron reflects faster moving acoustic phonons that impinge on it. In the complementary limit the large polaron is scattered by stationary optic phonons. That is, phonons' dispersion determines whether quasi-particles' relative motion is attributed primarily to phonon wavepackets or to large polarons. The scattering peaks when the wavelength of the faster quasi-particle coincides with the spatial extent of the self-trapped carrier. Beyond these idealized models self-trapped electrons generally interact with both acoustic and optic modes. Furthermore optic modes often possess significant dispersion.

11

Small-polaron transport

11.1 Loss of coherence and hopping transport

Incoherent transport results when the energy emitted or absorbed in moving a carrier exceeds the energy characterizing iso-energetic transfer. This section illustrates why small-polaron motion is usually incoherent. In summary, coherence is lost at high temperatures where atoms move classically when the change in the energy of a small-polaron's self-trapped carrier as it moves between sites is greater than its electronic transfer energy. Coherence is lost at low temperatures when a small-polaron's transfer energy is so reduced by the requirement that atoms' quantum-mechanically tunnel that it is overwhelmed by site-to-site variations of a small-polaron's energy.

Consider a small polaron formed with the Holstein short-range electron–phonon interaction. Then the requirement for a small-polaron's energetic stability in a cubic crystal becomes $F^2/2k > W/2$, where, as earlier, F represents the force associated with the short-range electron–phonon interaction, k is the displaced atoms' stiffness constant and W is the electronic bandwidth [cf. Eq. (2.5) and Section 4.4]. In particular, the carrier-induced shift of atoms' equilibrium positions by F/k lowers the self-trapped carrier's energy by $F(F/k) = F^2/k$ while increasing the atomic strain energy by $k(F/k)^2/2 = F^2/2k$.

For a self-trapped carrier to move the associated atoms must alter their positions. With classical atomic movement, the self-trapped carrier's energy must be brought into coincidence with what it would be upon its occupying a different site. The minimum-energy coincidence raises the electronic energy at the occupied site by $F(F/2k) = F^2/2k$, halfway toward liberating the self-trapped carrier, while creating an equivalent potential well at the carrier's would-be destination. Such small-polaron motion is incoherent when the increase of the electronic energy exceeds this motion's coherence energy t, the electronic transfer energy: $F^2/2k > t$. This condition is generally fulfilled because stable small-polaron formation insures that $F^2/2k > W/2 = zt$, where z is the number of nearest-neighbors in the tight-binding approximation.

The periodicity of an ideal lattice guarantees that a small polaron might still move coherently at low temperatures where atoms' classical motions are suppressed. Then a small-polaron's self-trapped carrier moves between sites as the associated atoms tunnel so as to remove atomic displacements about an initially occupied site and create equivalent atomic displacements about the destination site. The bandwidth associated with such iso-energetic motion is greatest in the adiabatic limit where an electronic carrier is always able to follow atomic movement. Then, as indicated previously [cf. Eq. (4.42)], the zero-temperature adiabatic small-polaron bandwidth is

$$W_{SP} = 2zh\omega\sqrt{\left(\frac{F^2/2k}{\hbar\omega}\right)}\exp\left(-\frac{F^2/2k}{\hbar\omega}\right). \qquad (11.1)$$

This bandwidth vanishes in the classical limit $\hbar \to 0$ where atomic tunneling is proscribed. Furthermore, as described in Section 10.2, bandwidths arising from atomic tunneling of harmonically vibrating atoms fall with increasing temperature. Thus, the small-polaron band is widest for adiabatic motion at zero temperature. Nonetheless, the small-polaron bandwidth of Eq. (10.1) is extremely small. For example, with $(F^2/2k)/\hbar\omega = 10$ the small-polaron bandwidth is about three orders of magnitude smaller than the displaced atoms' characteristic vibration energy $\hbar\omega$.

The small-polaron bandwidth, the coherence energy for this motion, is generally so small that it is less than the disorder potential resulting from defects and impurities. Indeed, even the potential drop between sites separated by 10 Å induced by an applied field of 10^4 V/cm, as applied in transit experiments, exceeds the small-polaron bandwidth. The applied field is itself then enough to obliterate coherent motion. The coherent motion of small bipolarons is even easier to destroy since having two carriers move together has the effect of replacing F by $2F$ in Eq. (11.1). Thus coherent small-polaronic motion is generally suppressed. As a result small polarons and small bipolarons are usually envisioned as moving incoherently by sequences of thermally assisted hops.

11.2 Two complementary types of hopping transport

Phonon-assisted electronic hopping is exemplified (1) by carriers' low-temperature hopping between shallow impurities in conventional covalent semiconductors and (2) by carriers' high-temperature hopping between cations in some transition-metal oxides. In both cases the hopping rates rise with increasing temperature. However, there are qualitative differences between these two examples. These distinctions define two complementary types of hopping transport: shallow-impurity hopping and small-polaron hopping.

Phonon-assisted hopping between localized electronic states is driven by their electron–phonon interactions. These interactions facilitate the absorption and emission of phonons that accompany a phonon-assisted jump. Electron–phonon interactions are most effective for phonons whose wavelengths exceed localized electronic-states' radii.

Since the radii of the electronic states associated with shallow impurities in conventional semiconductors R_i are much larger than the interatomic separation a, $R_i \gg a$, these states only interact with the small fraction of phonons whose wavelengths are exceptionally long. As a result large-radius electronic states have a weak net interaction with phonons [e.g. $(F^2/2k)(a/R_i)^3/\hbar\,\omega < 1$]. Each hop between such states is then dominated by the absorption or emission of just a single long-wavelength phonon (Miller and Abrahams, 1960; Emin, 1974; 1975a). The rate for a thermally assisted jump in which a single phonon is absorbed is then proportional to the Bose factor $1/[\exp(\Delta/\kappa T) - 1]$, where Δ, the energy of the absorbed phonon, equals the energy difference between the hop's two electronic states. For an acoustic-phonon-assisted jump the restriction of the electron–phonon interaction to long-wavelength phonons limits Δ to a small fraction of the Debye energy. Thus the low-temperature jump rate for a hop upward in energy is thermally activated with an activation energy that is much smaller than the Debye energy. Indeed, typical activation energies for shallow-impurity conduction are about 10^{-4} eV–10^{-3} eV (Miller and Abrahams, 1960). The Bose factor ensures that this jump rate will rise in proportion to the temperature for $\kappa T \gg \Delta$.

A hop downward in energy between large-radii shallow-impurity states primarily occurs with emission of only a single very low-energy acoustic phonon. Downward jumps between sites having a greater energy disparity are thereby precluded. Negative differential conductivity can even result if hopping transport is impeded by the inter-site potential drop generated by an applied electric field exceeding the highest energy acoustic phonon with which a shallow impurity can interact (Emin and Hart, 1985).

Small-polaron hopping refers to the phonon-assisted jumping of a carrier through severely localized (collapsed) states. The severe localization of these states enables them to interact strongly with phonons of nearly all wavelengths thereby yielding a strong net electron–phonon interaction [e.g. $(F^2/2k)/\hbar\omega \gg 1$]. As a result, a small-polaron jump usually involves the absorption and emission of multiple phonons. Indeed, even at low temperatures small-polaron hops typically involve far more than the minimal number of phonons required for energy conservation (Emin, 1974). As the temperature is raised above the characteristic phonon temperature the jump rate rises in an Arrhenius manner with an activation energy that generally exceeds the characteristic phonon energy. This high-temperature behavior distinguishes small-polaron hopping from that between electronic states which are weakly coupled to phonons.

Hopping-type impurity conduction in conventional semiconductors usually involves very long hops between large-radius states. In particular, hopping-type impurity conduction is reported at dopant concentrations below 2×10^{17} cm^{-3} in n-type silicon and below 6×10^{15} cm^{-3} in n-type germanium (Miller and Abrahams, 1960). The spherically averaged radii of ellipsoidal donor-state wavefunctions in n-type silicon and germanium are about 20 Å and 45 Å, respectively. Thus, the inter-impurity separations are an order of magnitude larger than the donor-state radii. In these situations the transfer energy linking the impurity centers is envisioned as the small perturbation that enables a hop to take place. In other words, these hops are viewed as *non-adiabatic*.

A different situation often prevails for small-polaron hopping in transition-metal oxides. In particular, the distances between nearest-neighbor transition-metal ions are just several times these cation radii. In these circumstances the electronic transfer energies can be too large to treat them simply as perturbations (Herring, 1961). The small-polaron jumps are then to be treated as *adiabatic* (Herring, 1961; Holstein 1959b; Emin and Holstein, 1969; Emin, 2008).

11.3 Overview of polaron hopping

A small-polaron hop may be regarded as a three-step process. A jump begins when, amidst atoms' thermal vibrations, extraordinarily large atomic displacements occur near a self-trapped carrier. Appropriate large-amplitude fluctuations enable the self-trapped carrier to then transfer between sites. Finally these transitory large-amplitude atomic displacements relax, dissipating energy to the vibrations of surrounding atoms.

Exchange of vibration energy between atoms, governed by the vibrations' dispersion, is crucial to phonon-assisted hopping. Effects of vibrations' dispersion on small-polaron hopping will be explicitly considered in Sections 11.4 through 11.7. However this section's overview of the physics of a small-polaron jump only implicitly presumes the presence of adequate vibrational dispersion.

Figure 11.1 illustrates the evolution of the processes which dominate the transfer of a self-trapped carrier as the temperature is increased (Emin, 1982). At low temperatures hopping is dominated by transfer processes that cost little in strain energy as atoms remain close to their equilibrium positions. In particular, process *a* of Fig. 11.1 depicts a self-trapped carrier being transferred as atoms tunnel from positions commensurate with its occupying its initial site to equivalent positions that

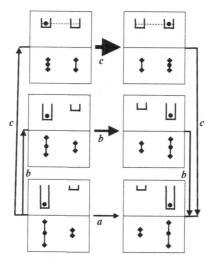

Fig. 11.1 A small-polaron hop occurs as a self-trapped carrier (denoted with a black dot) moves from a potential well centered on one site to an equivalent potential well centered on another (top subpanels) in response to changes of the atomic displacements at the two sites (bottom subpanels). The process requiring the smallest strain energy (process *a*) has the atomic displacements at the occupied site (schematically illustrated by a vertical line linking two diamonds) tunneling to values corresponding to the unoccupied site, while the atomic displacements at the initially unoccupied site tunnel to those for the occupied site. The speed of the transfer is indicated by the width of the central arrow which denotes it. Faster transfer occurs (in process *b*) when strain energy is first expended to alter the disparity of the configurations between which atoms tunnel. As depicted by process *c*, expending sufficient strain energy can eliminate the need for atomic tunneling between disparate atomic displacement patterns. The electronic carrier then transfers relatively rapidly between electronic states that are temporarily coincident.

correspond to the carrier being self-trapped at its final site. This atomic tunneling slows the transfer process. Thus, as indicated by arrow *a* in Fig. 11.2, the predominance of such atomic tunneling processes at low temperatures produces a relatively small jump rate.

Raising the temperature facilitates atoms assuming configurations requiring greater strain energy. As illustrated by process *b* in Fig. 11.1, atoms can then depart further from their equilibrium locations to positions that enable them to tunnel between less disparate configurations. Atomic tunneling's slowing of the transfer process is thereby reduced. As a result, depicted by arrow *b* in Fig. 11.2, the small-polaron hopping rate rises as the temperature is raised.

At high enough temperatures a small-polaron hop is dominated by transfers in which atoms surrounding initial and final sites are displaced sufficiently to bring the

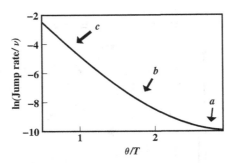

Fig. 11.2 A typical small-polaron jump rate in units of the characteristic atomic vibration frequency v is plotted against reciprocal temperature in units of the vibration temperature $\theta \equiv hv/\kappa$. At low temperatures the rate is very small and increases slowly with rising temperature as low-energy atomic-tunneling processes (e.g. process a of Fig. 11.1) predominate. The rate and its temperature dependence both increase with rising temperature as more efficient tunneling processes requiring larger strain energies (e.g. process b of Fig. 11.1) manifest themselves. At higher temperatures the rate becomes Arrhenius when transfer processes that eschew atomic tunneling but require relatively large strain energies prevail.

electronic energy of a carrier at the initial site into coincidence with what it would be if it occupied the final site. Then, as depicted by process c of Fig. 11.1, the carrier can move between sites without any atomic tunneling. The corresponding jump rate, indicted by arrow c in Fig. 11.2, then assumes an Arrhenius temperature dependence with its activation energy being the minimum strain energy required to form a coincidence. Since this activation energy only involves atoms' classical motion, it is independent of $\hbar$.

All told, Fig. 11.1 depicts increasingly energetic strained configurations enabling increasingly rapid electronic transfers. Electronic transfer is facilitated by reducing the associated atomic tunneling. The small-polaron hopping rate of Fig. 11.2 rises with increasing temperature as the probability of obtaining such energetic atomic configurations increases. The rate is non-Arrhenius at temperatures below that characterizing the energy of the phonons with which the self-trapped carrier inter-acts. At higher temperatures the rate becomes Arrhenius as atomic tunneling is supplanted by atoms' classical motion.

The ability of the carrier to move between sites in response to atomic motions depends on its electronic transfer energy. These electronic transfer energies are generally functions of atomic displacements. Indeed, the potential energies and the electronic wavefunctions that determine these electronic transfer integrals all depend on atoms' positions. Thus, an electronic transfer energy in a rigid solid differs from that governing a self-trapped carrier's optical transitions and from that for its semiclassical hopping (e.g. the electronic transfer energy at a coincidence

event). The dependence of a diffusing particle's transfer energy on atomic deformations can even be so strong that it qualitatively affects hopping diffusion (cf. light-
interstitial diffusion in metals, Chapter 17). Nonetheless, the electronic transfer
energy for a hop between sites g and $g + h$ is often approximated as the constant
one would obtain without polaron formation and any atomic displacements. In a
crystal this electronic transfer energy is just the Fourier transform of the carrier's
energy as a function of its quasi-momentum $E(k)$:

$$t(h) = \sum_k E(k)e^{ik \cdot h}. \tag{11.2}$$

This electronic transfer energy generally falls with increasing inter-site separation h
in a manner that reflects the complexity of the carrier's dispersion relation $E(k)$.
Nonetheless, transfer energies are often presumed to simply fall exponentially with
jump distance.

The simplest approach to calculating the small-polaron jump rate is to assume that
the electronic transfer energy is the small perturbation which enables a carrier to
move between sites. Section 11.4 describes a version of this (non-adiabatic)
approach that is generalized to address jumps between states that are inequivalent
in their energies and local electronic wavefunctions. Form factors permit these states
to have finite sizes and non-spherical shapes whereas simpler calculations implicitly
approximate the localized electronic states as just stationary points. The calculation
yields the full temperature-dependence of the non-adiabatic jump rate. In
Section 11.5 the high-temperature semiclassical limit of the non-adiabatic jump
rate is also derived using Holstein's occurrence-probability approach. However, the
non-adiabatic approach is only applicable if the electronic transfer energy is so small
that the carrier rarely transfers during a coincidence event. By contrast, the high-
temperature adiabatic regime discussed in Section 11.6 envisions the electronic
transfer being large enough (typically $t > \hbar \omega$) so that a carrier always hops at a
coincidence event. Two types of adiabatic jumps are discussed. In one scheme
coincidences are assumed to develop and relax rapidly (in less than a vibration
period). The second scheme envisions coincidences developing and relaxing slowly
as associated atomic vibrations possessing comparable frequencies are modulated
by inter-oscillator energy transfer. Section 11.7 describes how such weak vibrational
dispersion fosters temporal and spatial correlations between coincidence events.
These correlations cause a carrier's hops to occur in flurries separated by quiescent
periods. The semiclassical approach is then extended in Section 11.8 to address the
semiclassical hopping of a singlet bipolaron. It is shown that a singlet bipolaron's
two electronic carriers do not generally jump together as a pair. Rather, singlet-
bipolarons' semiclassical hopping is dominated by their thermally assisted

decomposition into separated polarons that hop independently of one another. Finally, Section 11.9 describes how polaron hopping in magnetic semiconductors is affected by their exchange interactions. In particular, ordering of a magnetic semiconductor's local moments alters its polarons' hopping conduction.

11.4 Non-adiabatic polaron jump rate

Non-adiabatic small-polaron hopping presumes very small electronic-transfer energies. Thus the self-trapped carrier usually remains localized and only occasionally avails itself of the opportunities to hop which atomic vibrations present to it. In this scheme the vibrations are approximated as just those about a carrier that is confined to a single site. For example, Fig. 11.3 plots that portion of the non-adiabatic (small-t) limit of the vibratory potential for two sites of Holstein's molecular crystal which depends on the two sites' relative displacement parameter $x_2 - x_1$:

$$V_r(x_2 - x_1) \equiv k(x_2 - x_1)^2/4 \pm F(x_2 - x_1)/2, \qquad (11.3)$$

where the $\pm$ sign pertains when sites 1 and 2 are, respectively, occupied. Here x_n is the displacement parameter for the n-th molecule. When this molecule is occupied by a carrier its equilibrium displacement is shifted by F/k. The darker curve is the relative potential when site 1 is occupied. Similarly the lighter curve is the relative potential when site 2 is occupied. A non-adiabatic hop from site 1 to site 2

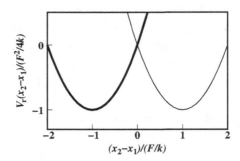

Fig. 11.3 This configuration diagram is often used to describe non-adiabatic hopping between a pair of sites. That portion of the deformational potential energy in the weak-dispersion limit of Holstein's molecular-crystal model, measured in units of $F^2/4k$, which depends on the relative atomic displacement coordinate of two adjacent sites $x_2 - x_1$ is plotted against this parameter, measured in units of F/k. The dark and light curves correspond to the carrier's occupations of sites 1 and 2, respectively. The energy of these curves' crossing point exceeds that of either curves' minimum by $F^2/4k$, the activation energy for non-adiabatic hopping in its high-temperature Arrhenius regime.

occasionally occurs when vibrations take the system through the crossing point at $x_1 = x_2$ or when atomic tunneling transfers the system through the potential barrier separating the two potential wells. Passage through the crossing point corresponds to process c of Figs. 11.1 and 11.2. Tunneling through the barrier represents processes a and b of Figs. 11.1 and 11.2.

The non-adiabatic hopping rate depends on the velocity with which an electronic carrier transfers between sites in response to atoms' movement. Since the set of localized electronic states between which a carrier hops is generally incomplete, the carrier's velocity operator is best expressed as $v \equiv i(H_e r - r H_e)/\hbar$, where H_e is the Hamiltonian describing the carrier's movement (Holstein and Friedman, 1968). This approach insures that the expectation value of the velocity equals the time derivative of the expectation value of the carrier's position. Diagonal matrix elements of a carrier's position between localized electronic states define its location and are unbounded: $\langle \phi_g | r | \phi_g \rangle = g$, where g denotes the centroid of a localized electronic function. By contrast, off-diagonal matrix elements of a carrier's position are bounded: $\langle \phi_{g'} | r | \phi_g \rangle \rightarrow 0$ as $|g' - g| \rightarrow \infty$. DC conduction's unbounded motion is described by a velocity operator that retains only diagonal matrix elements of the position operator: $\langle \phi_{g'} | v | \phi_g \rangle = i t_{g'g}(g - g')/\hbar$, where the transfer energy is defined by $t_{g'g} \equiv \langle \phi_{g'} | H_e | \phi_g \rangle$ (Holstein and Friedman, 1968). However, ac conduction from bound carriers' polarization currents generally requires inclusion of the non-diagonal matrix elements of the position operator. Even a crystal's regular array of potential wells produces bounded polarization-type contributions to the matrix elements of a carrier's position vector (Karplus and Luttinger, 1954). Nonetheless, these polarization contributions do not affect the dc transport phenomenon of Wannier–Stark localization (Anderson, 1963; Emin and Hart, 1987a).

Most generally, a phonon-assisted hop occurs when an electronic carrier moves between sites in response to atoms' vibrations. In the non-adiabatic regime the electronic velocity associated with transferring an electronic carrier between sites is presumed to be very small. Equivalently the electronic transfer energy $t_{g,g'}$ linking localized electronic states centered on sites g and g' is treated as the perturbation that enables the transition. The validity of this approach relies on the smallness of the electronic transfer energy.

Calculations of the rate for a non-adiabatic phonon-assisted transition commonly employ the simplification of ignoring carrier-induced shifts of atomic vibrations' frequencies. Then one of several equivalent expressions for this phonon-assisted jump rate becomes

$$R_{g,g'} = \left(\frac{t_{g,g'}}{\hbar} \right)^2 e^{-\Delta_{g,g'} / 2\kappa T} e^{-2S_{g,g'}} \int_{-\infty}^{\infty} d\tau \, e^{F_{g,g'}(\tau)} \cos(\Delta_{g,g'} \tau / \hbar), \qquad (11.4)$$

where $\Delta_{g,g'} \equiv E_{g'} - E_g$ is the non-zero difference of the energies of the hop's initial and final states (Yamashita and Kurosawa, 1958; Holstein, 1959b; Emin, 1974; 1975a; Bergeron and Emin, 1977). The other terms in this jump-rate expression are defined by:

$$2S_{g,g'} \equiv N^{-1}\sum_q \Gamma_{g,g'}(q)\gamma(q)\coth\left(\hbar\omega_q/2\kappa T\right) \qquad (11.5)$$

and

$$F_{g,g'}(\tau) \equiv N^{-1}\sum_q \Gamma_{g,g'}(q)\gamma(q)\csc h\left(\hbar\omega_q/2\kappa T\right)\cos\left(\omega_q\tau\right), \qquad (11.6)$$

where atoms are presumed to vibrate harmonically among N modes of frequency ω_q and wavevector q. The spatial correlation factor

$$\Gamma_{g,g'}(q) \equiv f_g^2 + f_{g'}^2 - 2f_g\,f_{g'}\cos[q\cdot(g - g')] \qquad (11.7)$$

indicates that vibrations whose wavelengths exceed the inter-site separation tend to be ineffective in producing a hop since they do not alter the energy difference between the two sites. Moreover, the form factor

$$f_g(q) \equiv \langle\phi_g|e^{iq\cdot(r-g)}|\phi_g\rangle \qquad (11.8)$$

indicates that a localized electronic state interacts most effectively with phonons whose wavelengths exceed its spatial extent. For example, the form factor for a hydrogenic wavefunction of radius b_g centered at site g is $f_g(q) = [1 + (qb_g/2)^2]^{-2}$. Finally, $\gamma(q)$ describes the carrier's electron–phonon interaction; in particular, for the Fröhlich long-range electron–phonon interaction:

$$\gamma_{LR}(q) \equiv \frac{\dfrac{e^2}{a}\left(\dfrac{1}{\varepsilon_\infty} - \dfrac{1}{\varepsilon_0}\right)\left(\dfrac{q_D}{q}\right)^2}{\hbar\omega_q}, \qquad (11.9)$$

where q_D is the Debye wavevector. The deformation-potential and molecular-crystal-model short-range electron–phonon interactions give:

$$\gamma_{DP}(q) \equiv \frac{(Zq_D)^2/2k}{\hbar\omega_q} \qquad (11.10)$$

and

$$\gamma_{\text{MCM}}(q) \equiv \frac{F^2/2k}{\hbar\omega_q},$$

(11.11)

respectively, where Z is the deformation potential energy. All told, the q-dependences of the arguments of the summations of Eqs. (11.5) and (11.6) are governed by: (1) the nature of the electron–phonon interaction, (2) the sizes and separation of the two electronic states involved in the hop, and (3) the dispersion relations of the involved phonons.

At high enough temperatures with sufficiently strong electron–phonon coupling and dispersion of the vibration frequencies, $F_{g,g'}(\tau)$ develops a large primary peak at $\tau = 0$ (Holstein, 1959b). The contribution from near this primary peak, rather than the sum of contributions from subsidiary peaks, will then dominate the integral of Eq. (11.4) if the vibration dispersion is three-dimensional (de Wit, 1968; Emin, 1975a).

The contribution to the τ-integration of Eq. (11.4) from the primary peak occurs for $\omega_q\tau \ll 1$. One may therefore write $F_{g,g'}(\tau) \approx F_{g,g'}(0) - G_{g,g'}\tau^2/2$ within the integrand with

$$G_{g,g'}(T) \equiv N^{-1}\sum_q \Gamma_{g,g'}(q)\gamma(q)\csc h(\hbar\omega_q/2\kappa T)\omega_q^2.$$

(11.12)

Upon performing the τ-integration the non-adiabatic small-polaron jump rate becomes

$$R_{g,g'} = \left(\frac{t_{g,g'}}{\hbar}\right)^2 \sqrt{\frac{2\pi}{G_{g,g'}}} e^{-\Delta_{g,g'}/2\kappa T} e^{-E_{g,g'}(T)/\kappa T} e^{-(\Delta_{g,g'}/\hbar)^2/2G_{g,g'}(T)},$$

(11.13)

where

$$E_{g,g'}(T)/\kappa T \equiv 2S_{g,g'} - F_{g,g'}(0) = N^{-1}\sum_q \Gamma_{g,g'}(q)\gamma(q)\tanh(\hbar\omega_q/4\kappa T).$$

(11.14)

As illustrated in Fig. 11.2, this jump rate's temperature dependence becomes Arrhenius at sufficiently high temperatures. Then $E_{g,g'}(T)$ approaches the temperature-independent value

$$E_{\text{A}}^{g,g'} \equiv N^{-1}\sum_q [\gamma(q)\hbar\omega_q]\Gamma_{g,g'}(q)/4$$

(11.15)

with

$$G_{g,g'}(T) \to 8\kappa T E_{\text{A}}^{g,g'}/\hbar^2.$$

(11.16)

As a result the semiclassical limit for the non-adiabatic jump rate becomes:

$$R_{g,g'} = \left(\frac{t_{g,g'}^2}{\hbar}\right)\sqrt{\frac{\pi}{4\kappa T E_{g,g'}}}\, e^{-\left(4E_{g,g'}+\Delta_{g,g'}\right)^2/16E_{g,g'}\kappa T}. \tag{11.17}$$

The semiclassical jump rate's activation energy generally rises with increasing jump distance. In particular, the dependence of the activation energy of the non-adiabatic semiclassical jump rate in Eq. (11.15) on jump distance enters through the correlation factor Eq. (11.7). At sufficiently large jump distances the activation energy of the non-adiabatic semiclassical jump rate approaches its maximum value:

$$E_A^{\max} \equiv N^{-1}\sum_q \left[\gamma(q)\hbar\omega_q\right]\left\{f_g^2(q)+f_{g'}^2(q)\right\}/4. \tag{11.18}$$

The approach of this activation energy to this maximum is controlled by the q-dependences of the factors within the summation of Eq. (11.15).

The dependences of the non-adiabatic semiclassical jump rate's activation energy on jump distance are illustrated in Fig. 11.4 for long-range and short-range models of the electron–phonon interaction. In these examples, hops are taken to be between two equivalent electronic states of radius b separated by the distance h. For ionic materials' Fröhlich long-range electron–phonon interaction the activation energy given by Eq. (11.15) rises gently with increasing h, roughly as $(1/b - 1/h)$ at large separations. This behavior occurs because short hops in ionic materials require smaller shifts of distant ions than do long hops. By contrast, as depicted in Fig. 11.4, a short-range electron–phonon interaction generally produces a relatively steep rise of the energy of Eq. (11.15) with increasing h. The short-range interaction of Holstein's molecular-crystal model (MCM) produces the most extreme example (Holstein, 1959a). In particular, in the limit of dispersion-less optic phonons the MCM gives $E_A^{g,g'} = E_A^{\max} = F^2/4k$ for $h \neq 0$ because $f_g(q) = f_{g'}(q) = 1$. That is, this model gives the maximum activation energy applying to all non-adiabatic hops. This artifact results because the MCM implicitly presumes that the localized electronic states between which a carrier hops have arbitrarily small radii, $b \to 0$; see the discussion following Eq. (4.18).

As described by Eq. (11.14) and illustrated in Fig. 11.2, a small-polaron jump rate passes from an Arrhenius regime to a non-Arrhenius regime as the temperature is lowered below a significant fraction of the temperature corresponding to the maximum energy of the phonons with which a carrier principally interacts. Equations (11.5) and (11.6) indicate that the specific non-Arrhenius temperature dependence of the jump rate depends on (1) the range of the electron–phonon interaction, (2) the dispersion relations of the involved phonons, and (3) the separation between a hop's two sites. These non-Arrhenius temperature dependences often closely resemble

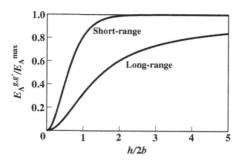

Fig. 11.4 The non-adiabatic hopping activation energy $E_A^{g,g'}$ in units of its maximum value E_A^{max} is plotted versus the jump length $h \equiv |g - g'|$ in units of twice the localized-state radius $2b$ for a hop between equivalent states $b_g = b_{g'} \equiv b$. The long-range electron–phonon interaction of an ionic medium produces a milder range-dependence than does a short-range electron–phonon interaction.

Upward transitions   **Downward transitions**

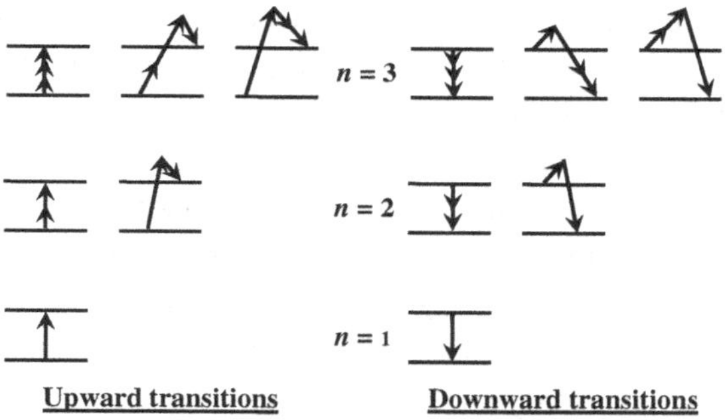

Fig. 11.5 Schematic description of processes in which the absorption and emission of one, two, and three acoustic phonons contribute to a transition upward or downward in energy.

those which are obtained by assuming percolative hopping with Arrhenius rates amongst energetically inequivalent sites (Emin, 1974).

Phonon-assisted transfers generally occur with multiple phonons being emitted and absorbed. Some processes involving the absorption of multiple acoustic phonons are illustrated in Fig. 11.5. In the extreme low-temperature limit when processes involving non-minimal absorption of phonon energy are completely frozen out, the processes depicted in the first columns of Fig. 11.5 persist. Then, since Eq. (11.6) gives $F_{g,g'}(\tau) \to 0$ as $T \to 0$, it is efficacious to expand $\exp[F_{g,g'}(\tau)]$ as a power series in the integrand of Eq. (11.4) to obtain:

$$R_{g,g'} = \left(\frac{t_{g,g'} e^{-S_{g,g'}}}{\hbar} \right)^2 e^{-\Delta_{g,g'}/2\kappa T} \sum_{n=1}^{\infty} \frac{1}{n!} \int_{-\infty}^{\infty} d\tau [F_{g,g'}(\tau)]^n \cos(\Delta_{g,g'}\tau/\hbar).$$

$$(11.19)$$

Here n, the power of $F_{g,g'}(\tau)$ in this series expansion, denotes the number of involved acoustic phonons and the τ-integration establishes energy conservation. That is, the net energy of the emitted or absorbed phonons equals $\Delta_{g,g'}$.

Evaluating the integrals for the deformation-potential interaction in the long-hop limit, $h \equiv |g - g'| \rightarrow \infty$, with $|\Delta_{g,g'}| < \hbar\omega_m$, where ω_m is the highest frequency acoustic phonon with which the localized carriers interact effectively, yields the low-temperature limit of the jump rate (Emin, 1974):

$$R_{g,g'} = \frac{2\pi \left(t_{g,g'} e^{-S_{g,g'}}\right)^2}{|\Delta_{g,g'}| \hbar} e^{-(\Delta_{g,g'} + |\Delta_{g,g'}|)/2\kappa T} \sum_{n=1} \frac{\left[6\gamma(\Delta_{g,g'}/\hbar\omega_D)^2\right]^n /(2n-1)!}{n!}$$

(11.20)

where $\gamma \equiv \gamma_{DP}(q_D)$ of Eq. (11.10). In this limit the jump rate for a hop downward in energy, $\Delta_{g,g'} < 0$, becomes temperature independent while the jump rate for a hop upwards in energy, $\Delta_{g,g'} > 0$, varies as $\exp(-\Delta_{g,g'}/\kappa T)$. In addition to Eq. (11.20) the general expression Eq. (11.4) indicates that these phonon-assisted jump rates satisfy the detailed-balance condition, $R_{g,g'}/R_{g',g} = \exp(-\Delta_{g,g'}/\kappa T)$.

When $|\Delta_{g,g'}| > \hbar\omega_m$, non-vanishing contributions to the summation begin at $n_m \equiv |\Delta_{g,g'}|/\hbar\omega_m$, the minimum number of acoustic phonons required by energy conservation. When $2n - 1$ is large enough Stirling's approximation can be applied to its factorial which is then combined with the expansion parameter to yield $\sim 6\gamma(\Delta_{g,g'}/n\hbar\omega_D)^2$. Analysis then shows that the downward jump rate falls once $|\Delta_{g,g'}|$ rises above $\sim\gamma\hbar\omega_m$. Similarly, the high-temperature jump rate for a hop downward in energy, Eq. (11.17), falls once $|\Delta_{g,g'}| > 4E_{g,g'}$. This feature indicates a bottleneck; the disorder energy becomes so large that the electron–phonon interaction is no longer strong enough to enable the multi-phonon emission associated with a downward transition to occur rapidly. As a result hopping downward in energy is suppressed when driven by a strong enough electric field (Emin, 1975a). This effect produces negative differential conductivity for hopping conduction (Emin and Hart, 1985). The occurrence of such bottlenecks also defines the "inverted" regime of chemistry's electron transfer reactions (Marcus, 1960).

11.5 Semiclassical treatment of the non-adiabatic polaron jump rate

The high-temperature jump rate in Eq. (11.17) is semiclassical in that it can be derived by treating atomic vibrations as classical (Holstein, 1959b). In this approach the electronic energy of a carrier on a site is taken to be a function of atomic positions that are themselves explicit functions of time τ. The jump amplitude $a_{g'}(\tau)$ for a carrier that begins on site g and moves to site g' under the influence of a small electronic transfer energy $t_{g,g'}$ is governed by the differential equation:

$$\left[i\hbar\frac{\partial}{\partial\tau}-E_{g'}(\tau)\right]a_{g'}(\tau)=t_{g,g'}\exp\left[-i\int^{\tau}d\tau'E_g(\tau')/\hbar\right], \tag{11.21}$$

where the exponential factor on the right-hand-side of Eq. (11.21) is the phase integral for a carrier confined to site g. Solving the differential equation to first order in $t_{g,g'}$ yields

$$a_{g'}(\tau)=\exp\left[-i\int^{\tau}d\tau'E_{g'}(\tau')/\hbar\right]\frac{t_{g,g'}}{i\hbar}\int d\tau\exp\left\{i\int^{\tau}d\tau'\left[E_{g'}(\tau')-E_g(\tau')\right]/\hbar\right\}. \tag{11.22}$$

The principal contributions to the second τ'-integration occur near points of stationary phase $\tau'=\tau_i$ defined by the coincidence condition $E_g(\tau_i)=E_{g'}(\tau_i)$. Near a point of stationary phase $E_g(\tau')-E_{g'}(\tau')\approx\dot{E}_{g,g'}(\tau'-\tau_i)$ with $\dot{E}_{g,g'}$ denoting the temporal rate of change of the energy difference $E_g=E_{g'}$ evaluated at $\tau'=\tau_i$. Using this approximation the second temporal integration is performed to obtain the transition amplitude associated with a particular coincidence event.

Most generally $a_{g'}(\tau)$ comprises a series of contributions from different coincidence events. The jump probability is the absolute square of this sum of transition amplitudes. However correlations between transition amplitudes from different coincidence events are eliminated with sufficient three-dimensional vibratory dispersion (de Wit, 1968; Emin, 1975a). The net probability of a jump is then the sum of the absolute squares of the jump amplitudes which are each associated with a coincidence event. The probability of a hop in response to a coincidence event characterized by $t_{g,g'}$ and $\dot{E}_{g,g'}$ is (Holstein, 1959b; Emin, 1991a):

$$P(\dot{E}_{g,g'})=\frac{2\pi\left|t_{g,g'}\right|^2}{\hbar\left|\dot{E}_{g,g'}\right|}. \tag{11.23}$$

This probability measures the ability of a carrier to move between sites during the time window presented to it by a coincidence event. That is, P is essentially the square of the product of the rate with which the carrier can move between sites having near equal electronic energies $|t_{g,g'}|/\hbar\equiv1/\tau_t$ and the duration of its coincidence $\tau_c\equiv(\hbar/\dot{E}_{g,g'})^{1/2}$. A coincidence's finite duration is a quantum-mechanical phenomenon: τ_c multiplied by the electronic-energy change during this time interval $|\dot{E}_{g,g'}|\tau_c$ satisfies the Heisenberg uncertainty relation, $|\dot{E}_{g,g'}|\tau_c^2\sim\hbar$ (Emin, 1973b).

Integrating the product of $P(\dot{E}_{g.g'})$ and the rate with which the corresponding coincidences occur over all thermally accessible values of $\dot{E}_{g.g'}$ yields the semiclassical jump-rate expression in Eq. (11.17) (Holstein, 1959b; Emin, 1991a). Its activation energy is simply the minimum strain energy required to establish a coincidence (Friedman and Holstein, 1963; Emin, 1973b). Thus the activation energy of Eq. (11.17) vanishes if the energy of the equilibrated carrier at g happens to equal that of the state at site g'.

For a hop to be non-adiabatic $P \ll 1$ its electronic transfer energy must be small enough: $|t_{g.g'}|^2 \ll \hbar |\dot{E}_{g.g'}|/2\pi \approx \hbar\omega(E_A \kappa T)^{1/2}$. Furthermore the semiclassical treatment requires $\hbar\omega(E_A)^{1/2} \ll (\kappa T)^{3/2}$ to assure that atoms' velocities do not change appreciably during a coincidence (Holstein, 1959b). Thus, a semiclassical non-adiabatic hop necessitates $|t_{g.g'}| \ll \kappa T$ (Emin and Holstein, 1969). The smallness of the electronic transfer energy implies a correspondingly small value for the pre-exponential factor in the expression for a non-adiabatic semiclassical polaron hop, Eq. (11.17). Thus, non-adiabatic semiclassical small-polaron hopping is characterized by very small values of the pre-exponential factors of its Arrhenius mobility and dc conductivity, $\ll 0.1$ cm^2/V-sec and $\ll 10^2/\Omega$-cm at 300 K, respectively. Observation of such small values might also indicate non-adiabatic hopping between defects and impurities since these hops are especially long with exceptionally small electronic transfer energies.

11.6 Semiclassical treatment of the adiabatic polaron jump rate

The rate for a phonon-assisted hop is generally larger in the adiabatic regime, where the electronic carrier moves rapidly enough to adjust to relevant atomic movement, than in the non-adiabatic regime, where the carrier is unable to follow such atomic motion. Nonetheless adiabatic and non-adiabatic hopping both manifest similar temperature dependences. As depicted in Fig. 11.2 low-temperature non-Arrhenius thermally assisted jump rates generally evolve to Arrhenius temperature dependences as the temperature is raised high enough for atoms' vibrations to be treated as classical. A hop in this Arrhenius semiclassical regime requires atomic displacements that bring the electronic energies of initial and final states into coincidence with one another. In other words, a semiclassical adiabatic hop requires establishing an atomic configuration for which the probabilities of a carrier occupying a hop's initial and final sites equal one another.

For the molecular-crystal model, adiabatic and non-adiabatic hops from site 1 to site 2 can be described with the aid of the anharmonic adiabatic potentials depicted in Fig. 11.6. Occupation of the left-hand well of lowest adiabatic potential corresponds to the carrier occupying site 1 while occupation of this adiabatic potential's right-hand well corresponds to the carrier occupying site 2. A semiclassical hop

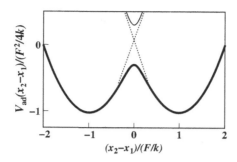

Fig. 11.6 The thick and thin solid curves are the lower and upper two-site adiabatic potentials of Eq. (3.14) plotted in units of $F^2/4k$ against the relative atomic displacement coordinate $x_2 - x_1$ in units of F/k for $F^2/4k = 3t$. Transitions between adiabatic potentials are depicted with dashed lines. The energy difference between either of the thick curve's minima and its maximum at $x_1 = x_2$ is the hopping activation energy $F^2/4k - t$. The lower adiabatic potential is strongly anharmonic in the vicinity of its maximum.

between sites 1 and 2 is described by trajectories which take a fictitious particle between the two wells of the lowest adiabatic potential. At a coincidence a trajectory can either remain on its adiabatic potential or make a non-adiabatic transition between adiabatic potentials. The trajectory for an adiabatic hop remains on the lowest adiabatic potential as it passes between its two wells. A trajectory for a non-adiabatic hop (1) transitions from the lower adiabatic potential to the upper adiabatic potential, (2) makes an odd number of oscillations through the upper adiabatic potential's minimum, and then (3) transitions from the upper adiabatic potential to the lower adiabatic potential. Even in the limit where non-adiabatic transitions predominate, a hop requires an adiabatic passage on either the upper or lower adiabatic potential. The non-adiabatic jump rate calculated in the preceding section is the rate for such an adiabatic passage when the electronic-transfer energy is arbitrarily small.

The net probability of a carrier transferring from site 1 to site 2, $W_t(1,2)$, is found by summing the probabilities of the direct adiabatic trajectory and all those involving non-adiabatic transitions (Holstein, 1959b):

$$W_t(1,2) = 2\frac{1 - e^{-(t_{1,2}/t_m)^2}}{2 - e^{-(t_{1,2}/t_m)^2}}, \tag{11.24}$$

where $t_m \equiv (\hbar|\dot{E}_{1,2}|/2\pi)^{1/2}$. Here $\exp[-(t_{1,2}/t_m)^2]$ is the probability of a non-adiabatic transition near the coincidence when its electronic transfer energy is $t_{1,2}$ (Zener, 1932). The plot of $W_t(1,2)$ versus $t_{1,2}/t_m$ in Fig. 11.7 shows that the non-adiabatic regime, where $W_t(1,2)$ is proportional to the square of $t_{1,2}$, is restricted to extremely

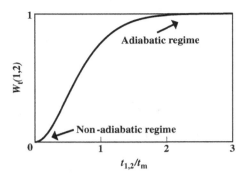

Fig. 11.7 The probability of a carrier moving from site 1 to site 2 in response to a coincidence increases with the electronic transfer energy $t_{1,2}$. Non-adiabatic hops, characterized by this transfer probability being proportional to the square of $t_{1,2}$, require extremely small transfer energies: $t_{1,2} \ll t_m$, where $t_m^2 \approx \hbar\omega(E_A\kappa T)^{1/2}$. The adiabatic regime is defined by this transfer probability approaching unity.

small values of the electronic transfer energy, $t_{1,2} \ll t_m$. The adiabatic regime, where $W_t(1,2)$ approaches unity, occurs for $t_{1,2} > t_m$. Estimates of nearest-neighbor electronic transfer energies frequently find them large enough, $> t_m \sim (\hbar\omega)^{1/2}(E_A\kappa T)^{1/4}$, for polaron hopping to be in the adiabatic regime (Herring, 1961). Nonetheless polaron hopping is often assumed to be in the non-adiabatic regime.

In the adiabatic limit, where a carrier always transfers between sites upon experiencing a coincidence, the potential governing jump-related atomic vibrations, the bold curve of Fig. 11.6, is anharmonic. By contrast, in the non-adiabatic limit, where a carrier rarely transfers between sites upon experiencing a coincidence, the potential governing jump-related atomic vibrations, the bold curve of Fig. 11.3, is harmonic.

Two complementary classes of semiclassical adiabatic polaron hop have been considered. These two types of adiabatic hop are represented by different kinds of trajectory that pass between the two minima of the lower anharmonic adiabatic potential of Fig. 11.6. In the weak-dispersion regime the amplitude of a trajectory that initially vibrates about one minimum gradually expands to transcend the energy barrier and encompass both minima. A trajectory that repeatedly transcends both wells and the negative-curvature region between them has its average frequency lowered (Emin, 2008). Ultimately after repeated passes the amplitudes of these double-well oscillations slowly diminish and the trajectory settles into oscillations about the other minimum. All told, the weak-dispersion regime envisions a hop's trajectory being slowly modulated by the transfer of energy between those atomic vibrations coupled to the carrier and other atomic vibrations. For example, energy transfer between vibrations of the molecules of Holstein's molecular-crystal model generates dispersion of the model's optic-vibration modes (Holstein, 1959b).

An adiabatic semiclassical hop in the strong-dispersion regime is represented by a ballistic-type of trajectory. Its trajectory suddenly rises from oscillations near the bottom of a potential well to a position above the energy barrier and then rapidly descends to oscillate about the double-well's other minimum. The hop is presumed to occur with no immediate return (Holstein, 1959b; Emin and Holstein, 1969). This presumption can be justified if a sufficiently large number of vibration modes having very disparate frequencies are strongly coupled to the carrier. Then phase coherence between these vibrations is lost within a typical vibration period. Thus interference obliterates the longer-lasting coherence characteristic of the weak-dispersion regime.

Several features of a semiclassical adiabatic hop enhance its rate above that of a non-adiabatic hop. First, as depicted in Fig. 11.7, the probability that a carrier will hop in response to a coincidence event is much larger for adiabatic hopping than for non-adiabatic hopping (Holstein, 1959b). Second, as illustrated in Fig. 11.6, transfer of a carrier in response to atomic movement lowers the peak of the adiabatic potential thereby reducing the hop's activation energy (Emin and Holstein, 1969). Third, in the weak-dispersion regime the reduction of the average vibration frequency for trans-barrier oscillations lowers these vibrations' free-energy. This effect augments the rate of slowly modulated coincidence formation (Emin, 2008).

Here the jump rate for an adiabatic semiclassical hop between degenerate sites is calculated for the weak-dispersion regime. The jump rate in the strong-dispersion regime is then obtained simply by neglecting the contribution to the weak-dispersion regime's adiabatic rate that comes from slowly modulated atomic oscillations. This result agrees with that obtained by directly calculating the rate for a semiclassical adiabatic hop in the strong-dispersion regime (Emin and Holstein, 1969).

In the adiabatic limit a carrier always follows atoms' motion. Concomitantly, as described in Chapter 4, these atoms' motions are also affected by a carrier's response to them. A polaron's adiabatic hopping rate is governed by the resulting atomic motion. At sufficiently high temperatures atoms' motion may be treated as classical. Then a hop requires atoms to pass through a coincidence configuration where the probabilities of the carrier occupying the jump's initial and final sites are equal. The semiclassical adiabatic jump rate generally has the form (Emin, 2008):

$$R_{ad} = v \exp[-(F_c - F_i)/\kappa T], \qquad (11.25)$$

where v is a representative atomic vibration frequency. Here F_c represents the system's free energy with atoms vibrating about the positions they assume during the hop's minimum-energy coincidence event. A coincidence is indicated by the $x_2 - x_1 = 0$ peak in Fig. 11.6 but appears as a saddle-point of the adiabatic potential when plotted against multiple deformation coordinates. The system's free energy with atoms vibrating about the displaced equilibrium positions they assume when the carrier occupies the hop's initial site is denoted by F_i.

Confining the self-trapped carrier at the hop's initial site shifts the equilibrium positions about which surrounding atoms vibrate. These atomic positions and associated stiffness constants are determined by minimizing the system's energy with respect to atomic displacements. The associated free energy F_i is the sum of energies arising from shifting atoms' equilibrium positions and the free energy of atoms' vibrations about these equilibrium positions:

$$F_i = \left(\frac{T_e}{R_i^2} - \frac{V_S}{R_i^3} - \frac{V_L}{R_i} \right) + f_i. \tag{11.26}$$

The lowering of the system's energy produced by shifts of atoms' equilibrium positions, the bracketed term of Eq. (11.26), depends on the parameters governing the formation of the self-trapped state of dimensionless radius R_i. These parameters, the carrier's kinetic energy T_e, its short-range electron–phonon interaction V_S and its long-range electron–phonon interaction V_L are defined in Eqs. (4.31)–(4.33). The self-trapped carrier's radius R_i minimizes the bracketed term of Eq. (11.26) subject to the structural constraints of a real solid. For example, the self-trapped carrier cannot be smaller than the relevant single structural unit: a molecule, one of its subunits, a bond or an ion (Emin, 2000a). The free energy associated with atoms harmonically vibrating about these displaced equilibrium positions is denoted by f_i.

At a coincidence the carrier is shared equally between initial and final sites (Emin, 2000a). A hop's characteristic coincidence is that which minimizes the coincidence energy with respect to atomic displacements (Friedman and Holstein, 1963). The free energy associated with atoms vibrating about the positions they assume at the minimum-energy coincidence is:

$$F_c = \left[\frac{T_e}{R_c^2} - \frac{V_S}{2R_c^3} - \frac{V_L}{2} \left(\frac{1}{R_c} + \frac{1}{S} \right) \right] - |t(S)| + f_c, \tag{11.27}$$

where the dimensionless separation between sites S greatly exceeds R_c, the dimensionless radius of each half of the shared self-trapped carrier (Emin, 2000a). Here the radius R_c minimizes the square-bracketed term of Eq. (11.27) subject to the solid's structural constraints (Emin, 2000a). The numerical coefficients of contributions to F_c that are explicitly related to electron–phonon interactions are half those of comparable terms in Eq. (11.26). This feature, reflective of the nonlinearity of self-trapping phenomena, occurs because dividing the self-trapped carrier into two equal parts reduces the polaron-related energy of each half by four: $2(1/4) = 1/2$. The

electronic carrier's expansion to the hop's two sites also lowers the carrier's confinement energy by $t(S)$, the electronic transfer energy linking the two sites. Finally,
f_c denotes the free energy of atomic vibrations associated with the coincidence
configuration.

Combining Eqs. (11.26) and (11.27) with Eq. (11.25) yields:

$$R_{ad} = v\exp[(f_i-f_c)/\kappa T]\exp(-E_a/\kappa T),\qquad(11.28)$$

where the activation energy is

$$E_a \equiv T_e\left(\frac{1}{R_c^2}-\frac{1}{R_i^2}\right)-V_S\left(\frac{1}{2R_c^3}-\frac{1}{R_i^3}\right)-V_L\left(\frac{1}{2R_c}+\frac{1}{2S}-\frac{1}{R_i}\right)-|t(S)|.\qquad(11.29)$$

Generally $R_c \geq R_i$ since the R_c-dependent electron–phonon interaction terms of Eq.
(11.27) are smaller than the analogous terms of Eq. (11.26) (Emin, 2000a).
Nonetheless, a common simplification is to set $R_c = R_i \equiv R$. Then the semiclassical
adiabatic hopping activation energy becomes

$$E_a = \frac{V_S}{2R^3}+\frac{V_L}{2}\left(\frac{1}{R}-\frac{1}{S}\right)-|t(S)|.\qquad(11.30)$$

This activation energy generally rises with increasing inter-site separation. By itself,
this effect fosters a carrier executing short hops.

Movement of an electronic carrier in response to atoms' vibrations generally
reduces the associated stiffness constants thereby decreasing the vibrations' free
energy (Emin, 2000a). This effect is especially large for a carrier's semiclassical
adiabatic hop (Emin, 2008). Near a coincidence saddle-point the carrier transfers
between sites with their atomic deformation energies changing by only the jump's
electronic transfer energy. As a result, the carrier-induced lowering of vibrations' free
energy which enters into Eq. (11.28) is dominated by that attributed to a coincidence.

The lowering of vibrations' free energy arising from a coincidence is addressed
by considering matrix elements of the electron–phonon interaction between the
closely spaced pair of electronic states associated with the coincidence. In particular,
evaluation of Eq. (4.37) for the even-symmetry lowest adiabatic electronic wavefunction $\Psi_+(r)$ and its odd-symmetry first excited state $\Psi_-(r)$ yields

$$\frac{f_i-f_c}{\kappa T} \cong \frac{\int dr\Psi_+^*(r)\Psi_-(r)\int dr'\Psi_+(r')\Psi_-^*(r')I(r,r')}{t_{1,2}}.\qquad(11.31)$$

Here $I(r,r')$ is the electron–phonon interaction function defined in Eq. (4.23) and the two adiabatic electronic wavefunctions are evaluated at the coincidence between sites 1 and 2 where their energies are separated by $2t_{1,2}$.

These two adiabatic electronic wavefunctions are then represented in terms of local electronic functions, $\phi_1(r)$ and $\phi_2(r)$, that, respectively, represent a carrier occupying sites 1 and 2:

$$\Psi_{\pm}(r) \equiv \frac{\phi_1(r) \pm \phi_2(r)}{\sqrt{2}}. \tag{11.32}$$

Writing $I(r,r')$ as the sum of its long-range and short-range components, Eqs. (4.27) and (4.28), and taking $\phi_1(r)$ and $\phi_2(r)$ as having identical forms albeit displaced from one another by the inter-site separation, Eq. (11.31) is evaluated to yield (Emin, 2008):

$$\frac{\Delta f^c_{vib}}{\kappa T} = \frac{E_a + t}{t}. \tag{11.33}$$

Feynman provides an alternative method of calculating the shift of vibrations' free energy produced by carrier-induced anharmonicity (Feynman, 1972; Emin, 2013).

Incorporating Eq. (11.33) into Eq. (11.28) yields the semiclassical adiabatic jump rate for the weak-dispersion model:

$$R_{ad} \cong \nu \left[\exp\left(\frac{E_a}{t}\right) \right] \exp\left(-\frac{E_a}{\kappa T}\right), \tag{11.34}$$

where $E_a \gg t \gg t_m \approx (\hbar\omega)^{1/2}(E_A\kappa T)^{1/4}$. The square-bracketed factor in Eq. (11.34) rises as the activation energy E_a is increased. This effect tends to offset the fall of the thermally activated factor with increasing activation energy. This feature is akin to the empirically observed Meyer–Neldel compensation effect (Meyer and Neldel, 1937).

The adiabatic jump rate in the strong-dispersion regime has been computed with the occurrence-probability approach. The occurrence-probability approach for ballistic trajectories ignores atoms' vibrations and identifies the adiabatic jump rate with the thermally averaged rate of coincidence formation with $R_{ad} \equiv \langle \dot{E}_{1,2} \rangle \langle \delta(E_1 - E_2) \rangle$, where the first average is over atomic velocities and the second average is over atomic positions (Holstein, 1959b; Emin and Holstein, 1969). The resulting rate is given by Eq. (11.28) but with $f_i - f_c = 0$.

Adiabatic hopping is distinguished from non-adiabatic hopping by the magnitude of the non-Arrhenius factor of its jump rate. In particular, in the weak-dispersion regime the square-bracketed factor of the adiabatic rate, Eq. (11.34), is larger than one. In the strong-dispersion regime the square-bracketed term of the adiabatic rate is just unity. By comparison the corresponding square-bracketed factor of the non-adiabatic rate is always very much less than one:

$$R_{\text{non-ad}} = \nu \left[\left(\frac{t}{t_{\text{m}}} \right)^2 \right] e^{-E_A/\kappa T} = \left(\frac{t^2}{\hbar} \right) \sqrt{\frac{\pi}{4\kappa T E_A}} e^{-E_A/\kappa T}, \qquad (11.35)$$

where $t \ll t_{\text{m}}$ and Eq. (11.17) has been rewritten with $\Delta_{g,g'} = 0$. Thus the magnitude of a jump rate's non-activated factor indicates whether the hop is adiabatic or non-adiabatic.

Several effects contribute to the small-polaron jump rate usually falling with increasing inter-site separation. First, the hopping activation energies generally rise with increasing inter-site separation. Even with the Meyer–Neldel compensation effect of Eq. (11.34) the adiabatic limit's jump rate falls with increasing inter-site separation provided that $t > \kappa T$. Second, as depicted in Fig. 11.7, the decrease of electronic transfer energies with increasing inter-site separation causes the jump rate to fall out of the adiabatic regime. Inter-site electronic transfer then becomes too sluggish for carriers to always avail themselves of the opportunities to hop presented by atoms being displaced into coincidence configurations. At large enough inter-site separations the electronic transfer energy becomes so small, $t \ll t_{\text{m}} \sim (\hbar\omega)^{1/2} (E_A\kappa T)^{1/4}$, that the jump rate falls into its non-adiabatic regime where it is proportional to the square of the electron-transfer energy (Holstein, 1959b).

Applying an electric field of magnitude F adds a contribution to the hopping activation energy. The added contribution lowers the activation energy for jumps facilitated by the field and raises the hopping activation energy for jumps in the opposite direction. In sufficiently strong fields the current flow becomes non-Ohmic as carriers overwhelmingly jump in concert with the applied field (Emin, 1975a).

The electric-field-dependent change of the potential energy of a carrier as it hops is typically smaller than the field-free activation energy. Then the electric field only contributes an additive correction to the hopping activation energy. This contribution is half the change of the carrier's potential energy garnered in the hop (Emin, 1975a). Thus the field-dependent contribution to a hop's activation energy is proportional to the jump's length.

As noted below Eq. (11.30), field-independent contributions to the activation energy also generally increase with a hop's length. With the long-range electron–phonon interaction of polar and ionic materials the jump-distance-dependent contribution to the hopping activation energy increases in proportion to the negative of the inverse jump distance. Such a hop's activation energy rises with increasing jump distance because greater shifting of distant ions is then required to produce a coincidence.

These two effects combine to contribute an explicit jump-length-dependent factor to the jump rate for a hop in concert with the applied electric field F (Emin, 2008):

$$\exp\left[\left(\frac{eFaS}{2} + \frac{V_L}{2S}\right)/\kappa T\right] = \exp\left\{\frac{\sqrt{eFaV_L}}{\kappa T}\left[1 + \frac{(S-S_0)^2}{2SS_0}\right]\right\}. \qquad (11.36)$$

Here S is the jump distance in units of the interatomic separation a, the term containing V_L comes from Eq. (11.30), and $S_0 \equiv (V_L/eFa)^{1/2} = [e(1/\varepsilon_\infty - 1/\varepsilon_0)/2F]^{1/2}/a$. Typically $S_0 \sim 10/a(\text{Å})$ for $F = 10^6$ V/cm, where $a(\text{Å})$ is the interatomic separation measured in angstroms.

Longer-range hops whose rates are suppressed by their relatively large field-free activation energies are enabled by the application of a sufficiently strong electric field. The factor in Eq. (11.36) becomes especially simple when the electric field's strength is large enough to lower S_0 to close to the suppressed hops' characteristic jump distance S: $(S - S_0)^2 < 2S_0^2$. The electric-field dependence of the jump rate then rises with electric-field strength as $\exp[CF^{1/2}/\kappa T]$ with $C \equiv [e^3(1/\varepsilon_\infty - 1/\varepsilon_0)/2]^{1/2}$. Measures of electric-field driven charge transport possessing this form are said to manifest Frenkel–Poole behavior (Frenkel, 1938).

11.7 Vibrational relaxation and correlations between polaron jumps

The transfer of energy between vibrating atoms is essential to phonon-assisted hopping. For a thermally activated semiclassical hop to occur, the atoms involved in forming a coincidence must absorb an extraordinary amount of energy from the vibrations of surrounding atoms. After the hop's charge transfer this vibrational energy is dissipated to surrounding atoms.

The transfer of energy between harmonically vibrating atoms is governed by the dispersion of their vibration frequencies. Holstein's molecular-crystal model envisions the carrier to be coupled to a single band of optic phonons (Holstein, 1959b). As such, it generally takes much longer than a typical optic-phonon vibration period $1/\omega$ to transfer a significant fraction of the strain energy utilized in forming a coincidence. In particular, the vibration energy remaining with atoms involved in a coincidence event produced with optic phonons after the time τ is $\sim E_A(\omega_w\tau)^{-d/2}$ for $\omega_w\tau > 1$, where ω_w is the width of the band of optic phonons and d is the vibrational system's dimensionality (Emin, 1970; 1971a; 1971b). The vibrational "hot spot" associated with a coincidence is deemed to have relaxed when its residual vibrational energy falls to κT. With $E_A \gg \kappa T$ and $\omega \gg \omega_w$ this relaxation time is much greater than $1/\omega$.

During relaxation from a coincidence the rates with which the involved sites form additional coincidences are enhanced. During a coincidence's relaxation there is also an increased probability that one of its sites will form a coincidence with a site that was not involved in the prior coincidence. In these instances the amount of energy that must be garnered from the vibrations of surrounding atoms to establish another coincidence rises as the energy utilized in forming the prior coincidence dissipates.

A carrier which responds adiabatically to two sites' recurring coincidences will hop back and forth between them so as to occupy the site of lowest electronic energy (Emin and Kriman, 1986; Emin, 1987). This situation is represented by trajectories that oscillate over the barrier of the adiabatic potential of Fig. 11.6. Furthermore a carrier that responds adiabatically to a coincidence with a fresh site will manifest an augmented jump rate prior to vibrational relaxation. These enhanced diffusion effects vanish in the extreme non-adiabatic limit where electronic transfer rates are much slower than coincidences' relaxation rates. A carrier's response is then too slow for it to benefit from the enhanced probability of successive coincidence events prior to vibrational relaxation.

When vibrational relaxation produces significant temporal correlations between the occurrences of successive coincidences a carrier's hops occur in flurries (Emin and Kriman, 1986; Emin, 1987). There are bursts of rapid hopping separated by long quiescent intervals. During a period of active hopping there are frequent return hops. A carrier's actual diffusion can occur when it abandons shuttling between two sites to begin back-and-forth motion between one of these sites and a fresh site. When in thermal equilibrium, active hopping and dormant periods compensate one another to yield the previously determined jump rates (Holstein, 1959b; Emin and Holstein, 1969; Emin, 2008). A series of Monte Carlo molecular dynamics calculations of the hopping diffusion of a vacancy in model substances report analogous correlation effects (Bennett and Alder, 1968; 1971; Bennett, 1975).

Polarons formed from carriers injected into a material do not begin in thermal equilibrium. Equilibration as polarons requires carriers' self-trapping with their stabilization energy being dissipated via vibrations' dispersion. Prior to their equili-bration as polarons injected carriers may move with greatly enhanced mobility. Relaxation can be so slow that transient high-field hopping conduction can be qualitatively different from polarons' steady-state hopping (Emin and Kriman, 1986).

Imagine for example a charge carrier injected into a narrow-band insulator alighting on a molecule of Holstein's molecular-crystal model. As depicted in the left-hand column of Fig. 11.8, the carrier's arrival shifts the equilibrium positions of the molecule's atoms. As the occupied molecule relaxes to its shifted equilibrium

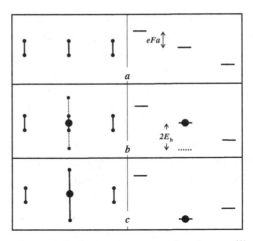

Fig. 11.8 Deformations and electronic energy levels are illustrated for three molecules of Holstein's molecular-crystal model with lattice constant a placed in electric field F. Panels a depict the carrier-free molecules at equilibrium. Panels b show the shift of the equilibrium deformation of a molecule (dashed line) and the corresponding lowering of its electronic energy $2E_b$ (dashed level) for $2E_b > eFa$ immediately after a carrier (the large dot) is placed on it. Panels c depict the equilibrated situation.

position vibrational energy comparable to Holstein's small-polaron binding energy E_b is transferred to surrounding molecules. The right-hand column of Fig. 11.8 indicates that the carrier's electronic state is not initially coincident with those of other molecules residing in the direction of an applied electric field F. However, the carrier's electronic energy passes through coincidences with molecules that reside in the down-field direction as the occupied molecule relaxes. If the carrier has a high probability of transferring in response to such a coincidence it will produce activation-less hopping with its mobility comparable to $ea^2v/\kappa T$, typically about 1 cm^2/V-sec (Emin and Kriman, 1986; Emin, 1987). This relatively rapid transient hopping motion will cease once a carrier lingers long enough at a site to permit lattice relaxation.

The mobility determined in a time-of-flight mobility measurement depends on the relationship between the experiment's transit time and the carriers' relaxation times. Transit-time measurements performed on thin films may be short enough to reveal only the carriers' unrelaxed mobility (Emin and Kriman, 1986). Even carriers' transits through bulk materials may not afford sufficient time for carriers' full equilibration. For example, the multiple captures and releases of carriers from deep traps required for complete equilibration may take longer than carriers' transits through bulk materials (Gibbons and Spear, 1966).

11.8 Pair-breaking in semiclassical singlet-bipolaron hopping

Singlet small-bipolaron formation (cf. Section 7.2) is often described with the Holstein–Hubbard model. Then two carriers with opposing spins that share a common site displace the equilibrium value of its atomic-deformation configuration coordinate by twice that as from a single localized carrier. Since each of the two carriers benefits from the energy lowering produced by the doubled atomic displacements the pair's net energy lowering is four times that for a solitary localized carrier $-E_b$. With U denoting the singlet pair's mutual Coulomb repulsion, the singlet bipolaron's net energy becomes $-4E_b + U$. Singlet-bipolaron formation is energetically favorable if this energy is less than the net polaron binding energy for two carriers localized at equivalent separated sites $-2E_b$. The small-bipolaron binding energy, the difference between these two energies, is $\varepsilon_b \equiv 2E_b - U$.

The four processes by which a singlet bipolaron can participate in charge transport are depicted in Fig. 11.9 (Emin, 1996). A bipolaron's two carriers can hop in unison, one of a bipolaron's two carriers can hop to another site, two polarons can merge to form a bipolaron, and a carrier can transfer from a bipolaron to a polaron.

The activation energies characterizing these types of semiclassical adiabatic hops have been calculated for the Holstein–Hubbard model (Emin, 1996). The activation energy for the joint hopping of the bipolaron's two carriers is $E_A(\text{joint}) \equiv 2E_b - t_2$, where the first contribution is simply four times the corresponding contribution for a polaron hop and t_2 is the magnitude of the two-electron transfer energy associated with moving a pair of electronic carriers. The activation energy for the bipolaron's separation into two polarons is $E_A(\text{separate}) \equiv (4E_b - U)^2/8E_b - t_1$, where t_1 is the magnitude of the conventional one-electron transfer energy and $t_1 > t_2$. The activation energy for uniting two polarons to form a bipolaron is $E_A(\text{unite}) \equiv U^2/8E_b - t_1$. As required by the detailed-balance condition the difference between the activation energies for separating a bipolaron into two polarons and that for merging two polarons into a bipolaron is just the bipolaron binding energy ε_b. Finally, the

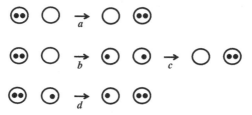

Fig. 11.9 Four transport processes involving a singlet bipolaron are (*a*) the joint transfer of a bipolaron's two carriers, (*b*) the transfer of one of a bipolaron's two carriers, (*c*) the uniting of two polarons into a bipolaron, and (*d*) the transfer of a carrier from a bipolaron to a polaron.

activation energy for transferring one of a bipolaron's two carriers to a polaron is found to be simply the activation energy for a single polaron hop, $E_A(\text{single}) \equiv E_b/2 - t_1$. With energetically stable bipolaron formation $\varepsilon_b > 0$, the activation energies for single-carrier transfers are smaller than that for the joint hopping of a bipolaron's two carriers: $E_A(\text{joint}) > E_A(\text{separate}) > E_A(\text{single}) > E_A(\text{unite})$.

Semiclassical dc hopping conductivity proceeds via the inter-site charge-transfer processes described above. The corresponding rates for these processes are: R (joint), $R(\text{separate})$, $R(\text{single})$ and $R(\text{unite})$. The dc conductivity is then (Emin, 1996):

$$\sigma = \frac{Ne^2a^2}{\kappa T}\left\{4f_b\, f_v[R(\text{joint}) + R(\text{separate})] + 2\, f_p(1-f_p)R(\text{single})\right\}, \quad (11.37)$$

where N is the site density and a is the nearest-neighbor separation. Here f_b and f_p respectively denote the fractions of sites occupied by bipolarons and polarons and f_v denotes the fraction of sites remaining vacant with $f_b + f_p + f_v = 1$. The numerical factors in Eq. (11.37) result because (1) $R(\text{joint})$ transports two charges rather than one, (2) each polaron state has a two-fold spin degeneracy, and (3) the contribution associated with merging polarons into bipolarons has been combined with that associated with separating bipolarons into polarons by using the detailed-balance relation $f_p f_p R(\text{unite}) = f_b f_v R(\text{separate})$.

The general expression for the dc conductivity of Eq. (11.37) can be culled to its most significant contribution. First, it is noted that $R(\text{separate}) > R(\text{joint})$ because $E_A(\text{separate}) < E_A(\text{joint})$. In addition, $R(\text{joint})$ will be reduced further if the two-electron transfer energy t_2 is small enough so that joint hops become non-adiabatic. Second, it is noted that $2f_pR(\text{single}) \gg 4f_bf_vR(\text{separate})$ in the physically relevant situation when most carriers form bipolarons: $f_b \approx f/2$, $f_v \approx 1 - f/2$ and $f_p \approx f^{1/2}\exp(-\varepsilon_b/2\kappa T)$, where f is the net number of charge carriers per site. In particular, the net activation energy for single-polaron hops, $\varepsilon_b/2 + E_A(\text{single})$, is considerably smaller than that for bipolaron's break-up, $E_A(\text{separate})$. Thus, the dc conductivity in the thermally activated semiclassical regime is dominated by polaron hops (Emin, 1996):

$$\sigma \cong \frac{2Ne^2a^2}{\kappa T}f_pR(\text{single}), \quad (11.38)$$

where the net activation energy is $3E_b/2 - U/2 - t_1$.

All told, polaron hopping becomes ever more predominant as E_b and the bipolaron binding energy ε_b are increased. In other words, as ever stronger binding increases the fraction of carriers that form bipolarons their hopping becomes so slow that the semiclassical dc conductivity is dominated by the relatively rapid motion of the fraction of carriers that remain as polarons.

11.9 Exchange interactions and polaron hopping in magnetic semiconductors

Polaron hopping has been reported to occur in numerous magnetic semiconductors. In these instances a polaron's inter-site motion is affected by interactions of its self-trapped carrier with the electrons that generate the material's local magnetic moments. The local magnetic moments are solely attributed to electrons' spins when, as is common, orbital magnetic moments are quenched by a material's crystal field. Then exchange interactions among valence electrons having unpaired spins govern the host material's magnetic interactions. A polaron's self-trapped carrier experiences the material's magnetism through its exchange interactions with valence electrons having unpaired spins.

When obeying Hund's rule the spin of a carrier localized at a site aligns with its valence electrons' spins to produce a net spin state of maximum multiplicity. The transfer of a carrier between two sites then depends on the relative alignment of these sites' magnetic moments. In the classical limit, in which the local moments are presumed to be arbitrarily large, the carrier's transfer energy becomes proportional to the spinor overlap factor: $\cos(\theta/2)$, where θ is the angle between the two sites' local moments. Then $\cos(\theta/2) = 1$ for parallel alignment $\theta = 0$ and $\cos(\theta/2) = 0$ for anti-parallel alignment $\theta = \pi$. Thus, the classical limit prohibits hops between sites whose magnetic moments are anti-parallel. However, this simple classical limit can be misleading. In particular, for finite-sized local moments the factor corresponding to $\cos(\theta/2)$ with $\theta = \pi$ becomes $1/(2S + 1)$, where S is the net spin of a site's valence electrons (Anderson and Hasegawa, 1955). Consider for example an electron hopping in a $S = \frac{1}{2}$ magnet or a hole hopping in a $S = 1$ magnet. Then, rather than being reduced to zero, the spinor overlap factor only falls to $\frac{1}{2}$ for electron transfer between sites with oppositely aligned $S = \frac{1}{2}$ valence spins. The lowering of the transfer energy arising from such a modest reduction of spinor overlap generally will not be sufficient to qualitatively alter a polaron's hopping rate.

The above discussion has modeled the transfer of an electron in a magnetic material as a one-electron process constrained by intra-site exchange interactions. However, exchange interactions result from indistinguishable electrons' mutual Coulomb repulsions. Thus electron transfer between magnetic sites is treatable as a multi-electron process in which one electron moves from a site containing $n + 1$ unpaired indistinguishable electrons to a site containing n unpaired indistinguishable electrons (Liu and Emin, 1979; Emin and Liu, 1983). The simplest example emerges when addressing the transfer of an electron from a site containing two electrons to a site containing one electron. For transitions between Hund's-rule states this approach regains the previously discussed spinor overlap factors. The multi-electron treatment also obtains overlap factors for transitions involving non-Hund's rule states. However, the transfer energy associated with transitions among Hund's rule states differs from its one-electron value in that it contains direct and exchange contributions involving Coulomb interactions between the carrier and valence electrons. Different combinations of these two-electron energies contribute to the transfer energies associated with non-Hund's rule states.

Magnetic ordering generally affects the activation energy for semiclassical polaron hopping. The most significant effect occurs with antiferromagnetic ordering. The activation energy for hopping between different sublattices of an antiferromagnet increases by about κT_N upon lowering the temperature below the Néel temperature T_N (Liu and Emin, 1979; Emin and Liu, 1983). The activation energy is augmented for these polaron hops because they occur between sites for which the ordered-state's internal magnetic field changes sign.

12

Polarons' Seebeck coefficients

12.1 Definitions and general concepts

The Seebeck coefficient is defined as the open-circuit emf developed across a material in response to an imposed temperature differential: $\alpha \equiv \Delta\phi/\Delta T$. A general result of irreversible thermodynamics is that a material's Seebeck coefficient is also the entropy transported with a carrier during isothermal current flow divided by the carrier's charge q (Callen, 1960). As a result, the Peltier heat Π, the heat transported with a charge carrier during isothermal current flow, is related to the Seebeck coefficient by $\Pi \equiv qT\alpha$. The third law of thermodynamics ensures that Π and α both vanish as the temperature is lowered toward absolute zero.

The entropy transported with a carrier is the sum of two components (Emin, 2002). The first component only depends on the carrier's presence whereas the second component depends on the details of the mechanism by which it moves. In particular, the first component is the change of the material's entropy that results from adding a charge carrier. The second component is the net energy transfer required to move a carrier divided by the temperature. The first component generally contributes to polarons' Seebeck coefficients. The second component can contribute significantly when polarons move by phonon-assisted hopping.

Since a polaron's motion requires atomic motion a crystal's polaron bandwidth is generally very narrow, less than the characteristic phonon energy. Polarons' Seebeck coefficients are therefore first considered in the limit of vanishing polaron bandwidth. The Seebeck coefficient arising from adding a polaron of energy E and chemical potential μ_c is then $\alpha = (E - \mu_c)/qT$. Neglecting explicit consideration of polaron spins this Seebeck coefficient can be rewritten in terms of the fraction of the energy band's states that are occupied, $f = 1/\{\exp[(E - \mu_c)/\kappa T] + 1\}$ as $\alpha = (\kappa/q)$ $\ln[(1 - f)/f]$. In this form it is evident that the narrow-band limit of the Seebeck coefficient changes sign at the band's half-filling ($f = \frac{1}{2}$ and $E = \mu_c$). In these examples the entropy entering the Seebeck coefficient is recognized as just the change of the entropy-of-mixing S produced by adding a carrier to $n = fN$ existing carriers distributed among N states: $\alpha q = \partial S/\partial n$, where

$$S = -Nk[f \ln (f) + (1-f)\ln (1-f)]. \qquad (12.1)$$

The spins of polarons' self-trapped electrons also contribute to their entropy and thereby to their Seebeck coefficients. In non-magnetic solids a self-trapped carrier's spin renders each individual polaron two-fold degenerate. As a result the polaron's Seebeck coefficient becomes $\alpha = (\kappa/q)\ln[2(1 - f)/f]$. By contrast, bipolarons produced as carriers self-trap as singlet pairs (1) generate no spin entropy, (2) double the electronic charge on each site and (3) halve the fraction of sites that are occupied to yield the Seebeck coefficient: $\alpha = (\kappa/2q)\ln[(2 - f)/f]$.

The two-fold spin degeneracy of a polaron with magnetic moment μ_B in a non-magnetic solid can be lifted with the application of a magnetic field H. The polaron's spin-entropy S_{sp} can be readily computed from the spins' free-energy:

$$S_{sp} \equiv \kappa\{\ln[2\cosh(\mu_B H/\kappa T)] - (\mu_B H/\kappa T)\tanh(\mu_B H/\kappa T)\}. \qquad (12.2)$$

This spin entropy decreases from $\kappa\ln(2)$ at $H = 0$ to zero as $\mu_B H/\kappa T \to \infty$. This fall-off becomes significant for strong magnetic fields >10 T at low temperatures < 10 K when the polaron magnetic moment is taken as equal to that of a free electron. The polaron's magnetic moment can be determined by measuring the decrease of the Seebeck coefficient's spin entropy with increasing magnetic field. A zero magnetic moment implies that the carrier is actually a singlet bipolaron.

The energies of a semiconductor's electrons generally shift with temperature. These energy shifts result from two effects that are both proportional to the net phonon energy. Electronic energy levels shift because (1) atoms' equilibrium positions shift with changing temperature and (2) semiconductors' energy bands undergo phonon-mediated thermally induced broadening. These effects combine to generate the temperature dependence of a semiconductor's optical gap. The interactions that produce these effects could also affect carriers' Seebeck coefficients. The Seebeck coefficient arising from these interactions has been evaluated (Emin, 1977b; 1984; 1985):

$$\alpha = \frac{(\langle E + E_{ph}\rangle_1 - \langle E_{ph}\rangle_0 - \mu_c)}{qT}, \qquad (12.3)$$

where μ_c denotes the chemical potential. The curved-bracketed expression is the net change of the energy of the system upon adding a charge carrier. The first contribution is the thermal average of the sum of a carrier's energy E, now a function of phonon frequencies and occupation numbers, and the phonon energy E_{ph}. The subscript 1 indicates that the thermal average is performed with vibration frequencies that have been modified by the added carrier's presence. The second contribution within the curved bracket denotes the thermal average of the phonon energy in the absence of the carrier (denoted by the subscript 0). As indicated by optical measurements the thermal average of a carrier's electronic energy $\langle E \rangle_1$ shifts with temperature. The carrier-induced change of the thermal average of the vibration energy $\langle E_{ph} \rangle_1 - \langle E_{ph} \rangle_0$ is also temperature dependent. The temperature dependences of both terms become strongest in the classical limit where κT exceeds the phonon energies. However when these two temperature dependences become strongest they also cancel one another. In other words the terms within the curved brackets show no net temperature dependence within the high-temperature regime. Thus thermal expansion and thermally induced phonon-mediated broadening of a semiconductor's energy bands do not significantly affect the curved-bracketed expression.

The narrow-band limit fails when the thermal energy κT falls below the bandwidth. Beyond the narrow-band limit a Boltzmann-equation-based analysis finds the Seebeck coefficient to be an average of the Seebeck coefficients from states of different energy weighted by $\sigma(E)$, each energy's contribution to the net electrical conductivity:

$$\alpha = \frac{1}{qT} \frac{\int dE \sigma(E)(E-\mu_c)}{\int dE \sigma(E)} \,. \tag{12.4}$$

Indeed this average extends over all energy bands that contribute to a semiconductor's electrical conductivity. Generally contributions from nearly empty conduction-band states are negative and those from nearly filled valence-band states are positive. As the temperature is lowered the chemical potential tends to become lodged among partially occupied states that reside within a semiconductor's intrinsic energy gap. With the electrical conductivity in the low-temperature limit presumed to be dominated by transport among states that are within κT of μ_c, the third law of thermodynamics is satisfied since then $\alpha \to 0$ as $T \to 0$.

12.2 Hopping polarons' Seebeck coefficients

The Seebeck coefficient has been calculated for carriers that move by phonon-assisted hopping. In particular, the open-circuit emf has been computed for a carrier that hops between sites that are each assumed to be coupled to independent phonon baths held at different temperatures (Emin, 1975b). The Seebeck coefficient for a carrier of charge q hopping between sites i and j depends on the energies of equilibrated carriers at these sites, E_i and E_j, and on measures of these sites' electron–phonon coupling strengths, Γ_i and Γ_j :

$$\alpha_{i,j} = \frac{E_i\left(\dfrac{\Gamma_j}{\Gamma_i + \Gamma_j}\right) + E_j\left(\dfrac{\Gamma_i}{\Gamma_i + \Gamma_j}\right) - \mu_c}{qT}, \tag{12.5}$$

where T denotes the mean temperature of the two sites and μ_c signifies the carrier's chemical potential. In the semiclassical limit in which atoms vibrate classically, $\Gamma_i/(\Gamma_i + \Gamma_j) = E_{p,i}/(E_{p,i} + E_{p,j})$, where $E_{p,i}$ and $E_{p,j}$ denote the corresponding sites' polaron binding energies. A site's polaron binding energy is the net energy lowering produced by atomic displacements when a carrier occupies the site. As such, a site's polaron binding energy contributes $-E_{p,i}$ to its net energy E_i. The polaron binding energy of a localized state generally decreases as its spatial extent is increased.

For hopping between equivalent ($\Gamma_i = \Gamma_j$) iso-energetic ($E_i = E_j = E$) sites the Seebeck coefficient becomes $\alpha_{i,j} = (E - \mu_c)/qT$. The Seebeck coefficient's characteristic energy $E_\alpha \equiv E - \mu_c$ is then just the activation energy for generating carriers of energy E in the narrow-band limit of a semiconductor. As such, at sufficiently high temperatures E_α differs from the activation energy of the crystal's dc electrical conductivity E_σ by the activation energy of a hopping polaron's semiclassical mobility, $E_\mu \equiv E_\sigma - E_\alpha$. For polaron hopping the semiclassical mobility's activation energy generally exceeds the characteristic phonon energy. Thus, polaron hopping is indicated if the electrical conductivity's activation energy E_σ exceeds E_α by at least the energy of a typical phonon.

When hopping occurs between non-degenerate ($E_i \neq E_j$) states whose electron–phonon couplings are inequivalent ($\Gamma_i \neq \Gamma_j$) some vibrational energy is transferred with the carrier as it hops. For example, consider hopping between a strongly coupled state of energy E_s with electron–phonon coupling Γ_s and a weakly coupled state of energy E_w with electron–phonon coupling Γ_w. Then with $\Gamma_s \gg \Gamma_w$ the Seebeck coefficient of Eq. (12.5) becomes $\alpha_{i,j} = (E_w - \mu_c)/qT$. That is, the energy which enters into the Seebeck coefficient is that of the state whose electron–phonon coupling is weakest. A localized electron's energy and coupling to phonons tends to increase with the severity of its localization. As illustrated in Fig. 12.1 a hop from a strong-coupling state to the higher energy weak-coupling state occurs with the absorption of vibration energy $\Delta \equiv E_w - E_s$ at the strong-coupling site. Similarly a hop from the weak-coupling state to the strong-coupling state occurs with the emission of vibration energy at the strong-coupling site. That is, the vibration energy Δ is transferred with the carrier as it hops.

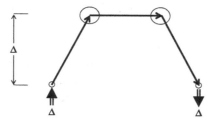

Fig. 12.1 Three hops are represented by solid arrows. The upward-directed double-lined arrow depicts absorption of vibration energy Δ at a severely localized state (small circle) facilitating a phonon-assisted hop to a weakly localized state (large circle). The horizontal solid arrow indicates a jump between equivalent degenerate weakly localized states. The downward-directed double-lined arrow depicts emission of vibration energy Δ at a severely localized state enabling a phonon-assisted hop from a weakly localized state.

In summary no vibration energy is transferred with direct polaron hops between equivalent states. As a result such polaron hopping yields a characteristic Seebeck energy that is less than the conductivity activation energy. By contrast, vibration energy is transferred with a carrier as it hops between non-degenerate inequivalent states. With this energy-transfer contribution the Seebeck energy becomes tied to the more weakly coupled state. Thus Seebeck coefficients of carriers moving among weak-coupling states are not affected by capture and release at strong-coupling traps. The conductivity activation energy and the characteristic Seebeck energy are then both associated with the weak-coupling states. Therefore comparison of the conductivity activation energy and the Seebeck energy provides a means of distinguishing between polaron hopping and trap-modulated (e.g. bandtail) conduction.

12.3 Softening (bi)polarons' Seebeck coefficients

Dynamic shifts of a localized carrier's charge distribution in response to atoms' vibrations generally alter their frequencies. In particular, the ground-state energy of a localized carrier is reduced as its centroid, size and/or shape adjust to the slowly modulating potential produced by vibrations of surrounding atoms. The localized carrier's relaxation thereby lowers the net energy and stiffness constants associated with these atoms' vibrational displacements. When a ground-state carrier's adjustment to atomic movement is treated by perturbation theory the lowered stiffness constants are given by Eq. (4.15) with the electron–phonon interaction inducing virtual transitions of the carrier from its ground state to its excited states.

As illustrated in Section 3.3, carrier-induced softening increases with the strength of the electron–phonon interaction until its matrix elements become large enough compared with the inter-level energy separations to induce a shift of atoms' equilibrium positions. Indeed, symmetry-breaking atomic displacements linking electronic states of different symmetry are maximally softened when their energy separation is just large enough to forestall the Jahn–Teller effect's symmetry-breaking shift of atoms' equilibrium positions (Jahn and Teller, 1937; Öpik and Pryce, 1957). A similar phenomenon affects a singlet bipolaron formed from degenerate orbital states. In particular, as discussed in Section 7.3, symmetry-breaking atomic displacements linking a singlet bipolaron's two-carrier states with different symmetries are most strongly softened when the difference of their exchange energies is just large enough to prevent symmetry-breaking shifts of atoms' equilibrium positions (Emin, 2000b). All told, carrier-induced softening becomes most significant when the carrier's electronic states are easily altered. Carrier's electronic states are exceptionally mutable when they are nearly degenerate or are large and particularly polarizable [cf. Eq. (4.18)].

Carrier-induced softening by a static polaron is absent from models whose simplifications preclude its localized carrier's wavefunction adjusting to movements of surrounding atoms. In particular, there is no carrier-induced softening for a stationary small polaron within Holstein's molecular-crystal model (Holstein, 1959b). Its small polaron is only characterized by shifts of the equilibrium values of its molecules' deformation coordinates.

Nonetheless, a softened region surrounding a localized carrier generally scatters phonons thereby increasing their thermal resistivity. Carrier-induced softening may even foster phonons' localization (Emin, 1990; 1991b). Moreover, carrier-induced softening enhances carriers' Seebeck coefficients via two effects (Emin, 1999). First, adding a charge carrier thereby (1) increases the entropy of atoms' vibrations and (2) augments the transfer of vibration energy associated with a polaron's hopping.

To illustrate these two effects consider polaron motion among molecules whose harmonic vibrations are softened upon occupation by a charge carrier. The entropy of an harmonic oscillator having angular frequency ω_i is $S_i = -\partial F_i/\partial T$, where the free-energy of the oscillator is $F_i = \kappa T \ln[2 \sinh(\hbar\omega_i/2\kappa T)]$. The first-order contribution to the Seebeck coefficient from a set of frequency shifts that are each denoted by $\Delta\omega_i$ is:

$$\alpha_{\text{vib}} = \frac{1}{q}\sum_i \frac{\partial S_i}{\partial \omega_i}\Delta\omega_i = \frac{\kappa}{q}\sum_i \left(\frac{-\Delta\omega_i}{\omega_i}\right)\left[\frac{(\hbar\omega_i/2\kappa T)}{\sinh(\hbar\omega_i/2\kappa T)}\right]^2. \tag{12.6}$$

As the temperature is raised from zero the square-bracketed expression increases monotonically from zero toward unity as $\hbar\omega_i/2\kappa T \to 0$.

Carrier-induced softening enhances the transfer of vibration energy with a thermally assisted hop. In particular, carrier-induced stiffness changes enable vibration energy to be transported with a carrier as it hops between equivalent sites. By contrast, in the absence of carrier-induced stiffness changes the transfer of vibration energy during a thermally assisted hop only occurs for hops between sites with inequivalent couplings to the phonons, cf. Section 12.2.

An example illustrates how a carrier-induced change of stiffness affects the transport of vibration energy. Consider the net harmonic deformation energy of two sites whose deformation coordinates are x_1 and x_2. A carrier's occupation of site 1 causes its deformation coordinate to be displaced to F/k_c as its stiffness constant is shifted from k to k_c:

$$E_{\text{def}} = \frac{k_c}{2}(x_1 - F/k_c)^2 + \frac{k}{2}x_2^2, \tag{12.7}$$

where F is the force associated with the electron–phonon interaction. A semiclassical hop between equivalent sites can occur when their electronic energies come into coincidence, $W(x_1) = W(x_2)$ at $x_1 = x_2$. The minimum deformation energy needed to establish such a coincidence occurs when $x_1 = x_2 = F/(k_c + k)$. The deformation energies at sites 1 and 2 associated with establishing this minimum energy coincidence are $F^2[k^2/k_c(k + k_c)^2]/2$ and $F^2[k/(k + k_c)^2]/2$, respectively. The sum of these two energies is the activation energy for the semiclassical polaron hop $E_b[k/(k + k_c)]$, where $E_b \equiv F^2/2k_c$ is the polaron's binding energy. The difference between the two sites' deformation energies at the minimum-energy coincidence characterizes the vibration energy transported in a hop, $E_b[k/(k + k_c)][(k - k_c)/(k + k_c)]$. To first order in the carrier-induced shift of the local vibration frequency $\Delta\omega \equiv \omega_c - \omega$ the transported energy becomes $(E_b/2)(-\Delta\omega/\omega)$, where the stiffness k is related to the vibration frequency ω by $k \propto \omega^2$.

This transported vibration energy divided by qT contributes to the Seebeck coefficient for high-temperature semiclassical polaron hopping. As the temperature is lowered below the high-temperature semiclassical limit the hopping activation energy and the vibration energy transported with a hop both decline. Indeed a general treatment of the vibration energy contribution to the Seebeck coefficient yields (Emin, 1999):

$$\alpha_{\text{trans}} = \frac{\kappa}{q}\sum_i \left(\frac{-\Delta\omega_i}{\omega_i}\right)\left(\frac{E_{b,i}}{2\hbar\omega_i}\right)\left[\frac{(\hbar\omega_i/\kappa T)^2}{\sinh(\hbar\omega_i/\kappa T)}\right] \tag{12.8}$$

for carrier-induced softening between equivalent sites, where $E_{b,i}$ is the contribution to the polaron binding energy from the i-th vibration mode. The square-bracketed factor in Eq. (12.8) peaks at $T \approx \hbar\omega_i/2\kappa$ while it vanishes as $T \to 0$ and as $T \to \infty$.

The enhancement of the Seebeck coefficient from carrier-induced softening is the sum of α_{vib} and α_{trans}. The temperature dependence of each of these contributions is governed by those of the square-bracketed factors contained within its i-summation. These factors, respectively defined as $f_{vib,i}$ and $f_{trans,i}$, are plotted in Fig. 12.2.

The magnitude of α_{trans} can be crudely underestimated by setting $E_{b,i}/2\hbar\omega_i = 1$. Carrier-induced softening's contribution to the Seebeck coefficient is then roughly κ/q multiplied by the product of the fractional softening $\Delta\omega/\omega$ and the number of softened modes. Typical carrier-induced softening is about 10% (Austin and Mott, 1969). Then each softened mode contributes ~10 μV/K to the net Seebeck coefficient.

However, as described at the beginning of this section, carrier-induced softening sometimes can be especially large. The unexpectedly large Seebeck coefficients in boron carbides have been attributed to especially large carrier-induced softening (Aselage *et al.*, 1998; 2001). This exceptional softening has been ascribed to the formation of singlet bipolarons amongst the nearly degenerate frontier orbitals of boron carbides' icosahedra (Emin, 2000b). A significant contribution to pentacene's Seebeck coefficient has also been attributed to carrier-induced stiffness changes (von Mühlenen *et al.*, 2007).

Semiconductors whose carriers yield unusually large Seebeck coefficients while also bolstering a material's thermal resistivity ρ_T can be useful in thermoelectric devices. Indeed, both effects enhance a material's thermoelectric figure-of-merit $\alpha^2\sigma\rho_T$, where σ denotes the electric conductivity. To this end pursuing materials that might have exceptionally large carrier-induced stiffness changes provides an avenue

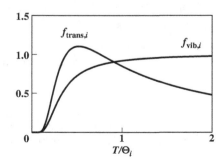

Fig. 12.2 The temperature-dependent factors $f_{trans,i}$ and $f_{vib,i}$ are plotted against T/Θ_i, where $\Theta_i \equiv \hbar\omega_i/\kappa$.

for finding promising materials for thermoelectric applications. For example, boron carbides' exceptional thermoelectric properties combined with the huge neutron absorption cross-section of ^{10}B provides the basis of a solid-state neutron detector (Emin and Aselage, 2005).

12.4 Seebeck coefficients of polarons in magnetic semiconductors

Polarons are reported in narrow-band magnetic materials including some transition-metal and rare-earth oxides. These materials' magnetism is attributed to local magnetic moments primarily from unpaired spins of electrons localized on cation sites. Intra-site exchange interactions foster alignment of each site's unpaired spins. In addition, comparatively weak inter-site exchange effects promote ordering of local moments' relative orientations. Super-exchange interactions involving the nominally non-magnetic ligands that reside between magnetic ions generate the antiferromagnetic orderings which characterize most magnetic insulators (Kramers, 1934; Anderson, 1950). The prototypical illustration is MnO, an antiferromagnetic insulator with a spin of 5/2 on each of its Mn^{2+} cations. Nonetheless direct exchange between magnetic ions sometimes generates ferromagnetic ordering. The classic example is EuO, a ferromagnetic insulator with a 7/2 spin on each of its Eu^{2+} cations (Methfessel and Mattis, 1968).

Adding a severely localized charge carrier to a magnetic site alters its magnetic moment and the associated exchange interactions. These changes alter the orientations available to the localized electrons. The magnetic system's entropy is thereby altered by the addition of a charge carrier. The carrier-induced change of the magnetic entropy divided by the carrier's charge provides the magnetic contribution to the carrier's Seebeck coefficient (Liu and Emin, 1984).

Local moments are well ordered at temperatures well below the magnetic ordering temperature T_m. Then the magnetic contribution to a carrier's Seebeck coefficient at temperature T is:

$$\alpha_{mag} \cong \mp \frac{\kappa}{q} \left[\left(\frac{T_m}{T} \right) \exp \left(-\frac{T_m}{T} \right) \right], \qquad (12.9)$$

where the upper and lower signs respectively pertain for a carrier increasing or decreasing the occupied site's magnetic moment (Liu and Emin, 1984). The magnitude of α_{mag} is very small at low temperatures and increases as the temperature is raised toward the magnetic ordering temperature.

At much higher temperatures $T > T_m$ the magnet enters its paramagnetic regime. As the temperature is increased constraints on the orientations of local moments are

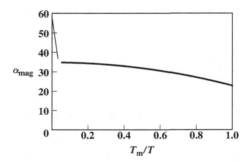

Fig. 12.3 The magnetic contribution to the Seebeck coefficient in the paramagnetic regime $T > T_m$ is plotted against T_m/T. The dark line is obtained from Eq. (12.10) with $S_c = 1$, $S = \frac{1}{2}$ and $A_{ex} = 5/36$. The thin curve indicates the behavior as the thermal energy is raised above the intra-site exchange energy.

reduced and they become free to rotate independently. The magnetic contribution to the Seebeck coefficient then takes the form (Liu and Emin, 1984)

$$\alpha_{mag} \cong \frac{\kappa}{q} \left[\ln \left(\frac{2S_c + 1}{2S + 1} \right) \mp A_{ex} \left(\frac{T_m}{T} \right)^2 \right], \tag{12.10}$$

where S_c and S are the values of the local moment's spin with and without the carrier's presence. The corresponding multiplicities of the local moment's spin state are $2S_c + 1$ and $2S + 1$. Again the upper and lower signs respectively pertain when $S_c > S$ and $S_c < S$ with A_{ex} denoting a numerical constant that is of the order of unity. In the extreme paramagnetic limit $T \gg T_m$ inter-site exchange becomes ignorable. Then only the first term within the square brackets of Eq. (12.10) survives (Heikes *et al.*, 1963). For large spins $S \gg 1$, $\alpha_{mag} \to \pm (\kappa/q)[1/(2S + 1)]$, where the sign indicates whether the addition of a carrier increases or decreases the site's spin.

Finally, in the generally unattainable ultra-high-temperature limit the thermal energy rises well above the intra-site exchange energy. In this regime even intra-site exchange effects can be ignored. Since the carrier is then unaffected by exchange interactions, the magnetic contribution to its Seebeck coefficient reverts to that for a spin-1/2 carrier in a non-magnetic solid: $\alpha_{mag} \to \alpha_{sp}$ of Eq. (12.2).

Figure 12.3 illustrates the results of a model calculation of α_{mag} in the paramagnetic regime (Liu and Emin, 1984). Two regions with different temperature dependences are encountered as the temperature is raised above T_m. These two regions correspond to the carrier's inter-site and intra-site exchange interactions being successively overwhelmed. The carrier-related spin entropy rises as the constraints imposed by these exchange interactions are progressively overcome.

13

Polarons' Hall effect

As illustrated in Fig. 13.1, a Hall Effect results when a magnetic field H applied perpendicular to the current density J induces development of a Hall electric field F_H in the mutually perpendicular direction. These measurable parameters can be combined to yield the Hall coefficient $R_H \equiv F_H/JH$.

The transverse component of velocity $v_\perp \equiv \mu_H v H$ imparted to a carrier having velocity v by the magnetic field H defines the Hall mobility μ_H. The Hall angle $\theta_H \equiv v_\perp/v = \mu_H v H/v = \mu_H H$ describes this deflection. The Hall angle and Hall mobility are positive (indicating counterclockwise flow) for classical electrons. However, there can be no net dc steady-state carrier flow in a Hall measurement's transverse direction. Rather, space charge develops in the transverse direction so as to generate the Hall electric field. The presence of the Hall electric field rotates the net electric field thereby compensating the deflection of the current by the applied magnetic field: $F_H/(J/\sigma) = -\mu_H H$, where σ is the electrical conductivity. Inserting the definition of the Hall coefficient R_H yields an expression for the Hall mobility: $\mu_H = -\sigma R_H$. Thus the Hall mobility is obtained from measurements of σ and R_H.

Writing the electrical conductivity as the product of the carrier's density n, charge q and mobility μ, $\sigma = nq\mu$, enables the Hall coefficient to be expressed as $R_H = (1/nq) (-\mu_H/\mu)$. Here μ_H and μ are recognized as different measures of carriers' intrinsic (trap-free) motion. The Hall mobility μ_H measures the magnetic-field-induced deflection of carriers' flow. By contrast, the standard mobility μ measures carriers' movement in response to an electric field. The standard mobility has also been called the "conductivity mobility" and the "drift mobility". But the term "drift mobility" also refers to the trap-limited mobility measured in time-of-flight experiments. Since $-\mu_H/\mu$ is positive for quasi-classical carrier motion, the sign of its Hall coefficient $R_H = (1/nq)(-\mu_H/\mu)$ gives the sign of the carrier charge q. Measurement of R_H for quasi-classical carriers also provides an estimate of the carrier density n since then $\mu_H \approx -\mu$ and $R_H \approx 1/nq$.

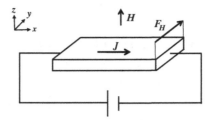

Fig. 13.1 The Hall electric field F_H develops across a material when the magnetic field H is applied perpendicular to the direction in which the current density J flows. For electrons' classical motion the Hall electric field is directed in the positive y-direction.

Polaron motion is usually qualitatively different from quasi-classical carrier motion. As a result, the magnitudes, temperature dependences and even the signs of polarons' Hall mobility are distinct from those of the quasi-classical carriers found in common semiconductors. Indeed, observing the distinctive features of polarons' Hall mobilities often facilitates identifying the carriers as polarons.

13.1 Coherent transport's Hall mobility

The Hall mobility can be calculated for coherent motion by solving the Boltzmann equation to first order in the strengths of both the electric and magnetic fields for the arrangement depicted in Fig. 13.1:

$$\mu_H = \frac{-q \sum_{k_x,k_y} f_0(1-f_0)\left[\left(\dfrac{v_y^2}{m_{xx}} - \dfrac{v_x v_y}{m_{xy}}\right)\tau^2\right]}{\sum_{k_x,k_y} f_0(1-f_0)\left(v_x^2\tau\right)}. \tag{13.1}$$

Here the distribution function for an equilibrated carrier of energy E is just the Fermi factor $f_0 = 1/\{\exp[(E - \mu_c)/\kappa T] + 1\}$, where μ_c denotes the carriers' chemical potential. The components of a carrier's velocity in the x- and y-directions, the partial derivatives of E with respect to the corresponding quasi-momentum $\hbar k_x$ or $\hbar k_y$, are denoted by v_x and v_y. Similarly, m_{xx} and m_{xy} are the components of the carrier's effective-mass tensor, the reciprocals of the second partial derivatives of E with respect to components of the carrier quasi-momentum. The carrier relaxation time is denoted by τ. All of these quantities, E, v_x, v_y, m_{xx}, m_{xy} and τ, are generally functions of k_x and k_y.

The Hall mobility, measuring the deflection of a current by a magnetic field, generally differs from the standard mobility, measuring the establishment of a current by an electric field. However, the magnitudes of these mobilities become comparable to one another in a semiconductor, $f_0 (1 - f_0) \to f_0 \approx \exp[-|E - \mu_c|/\kappa T] \ll 1$, with transport in a band that is much wider than κT. For example, $\mu_H = -\mu = -q\tau/m^*$ for charge transport involving states near a band extremum for which $|E| = m^*(v_x^2 + v_y^2 + v_z^2)/2$ when τ is approximated as a constant. Then the Hall coefficient becomes just $R_H = 1/nq$, where n is the density of mobile, untrapped, carriers.

A very different situation prevails if scattering-type transport involves an energy band whose width is considerably less than the thermal energy κT. In this limit f_0 becomes nearly independent of a carrier's wavevector. The mobility then reduces to

$$\mu = \frac{q}{\kappa T} \sum_{k_x} v_x^2 \tau. \tag{13.2}$$

Since each velocity component is proportional to corresponding bandwidth parameters the mobility is second order in them. Similarly, to second order in bandwidth parameters the Hall mobility of Eq. (13.1) becomes

$$\mu_H = \frac{-q \sum\limits_{k_x, k_y} \left[\left(\dfrac{v_y^2}{m_{xx}} - \dfrac{v_x v_y}{m_{xy}} \right) \tau^2 \right] \left(1 - \dfrac{E}{\kappa T} \right)}{\sum\limits_{k_x, k_y} \left(v_x^2 \tau \right)}, \tag{13.3}$$

where reciprocals of components of the effective mass tensor have been recognized as also being proportional to bandwidth parameters. In particular, the square-bracketed factor in the numerator is of third order in the bandwidth parameters while the denominator is of second order in them. The factor of E in the numerator raises it to fourth order in the bandwidth.

The temperature-independent contribution to the narrow-band limit of μ_H tends to vanish for the symmetric energy bands of highly symmetric structures. Thus, for example, a carrier's Hall mobility and the standard mobility are both temperature dependent in the narrow-band limit for square and face-centered cubic structures: $\mu_H = -\mu \propto 1/\kappa T$ when τ is assumed to be independent of the wavevector (Friedman, 1963). By contrast, the narrow-band limit for a carrier in a triangular lattice yields a Hall mobility that is temperature independent, larger than μ by the ratio of κT to the bandwidth, and can be anomalously signed (Friedman, 1963). The Hall constant then departs from its nominal value $1/nq$ in temperature dependence, magnitude and sign. Hall-effect sign anomalies occur when the contributions from states of negative effective mass dominate the summations of Eq. (13.3). For example, Hall-effect sign anomalies are found for coherent motion involving the narrow energy bands that are presumed to characterize anthracene's electronic transport (Friedman, 1964).

The Hall mobility of Eq. (13.1) applies to polarons whose inter-site motion is best described as being coherent but limited by occasional scattering events. As discussed in Section 10.1 large-polaron motion is generally coherent albeit with a narrow energy band. Such a large polaron's Hall mobility is described by Eq. (13.3).

This coherent transport picture has also been applied to a small polaron's motion in an ideal crystal at sufficiently low temperatures (Holstein, 1959b). With rising temperature the width of a small polaron's very narrow band decreases further and phonon scattering increases until coherent motion is destroyed. These two effects combine to preclude small-polarons' coherent motion in even an ideal crystal at all but low temperatures $\kappa T \ll \hbar\omega$ (Holstein and Friedman, 1968). Furthermore, as discussed in Section 11.1, coherent small-polaron motion is destroyed when even mild energetic disorder or the application of a modest electric field produces site-to-site energy differences that exceed the width of the extremely narrow small polaron band. For example, the small-polaron bandwidth for Holstein's molecular-crystal model is $\leq 6\hbar\omega\exp[-(E_b/\hbar\omega)\coth(\hbar\omega/2\kappa T)]$, where E_b is the small-polaron binding energy and $E_b \gg \hbar\omega$, cf. Section 4.4. Thus, in principle the small-polaron Hall mobility could be described with the narrow-band limit of Eq. (13.3) if the material is extremely well ordered, the temperature is very low and an applied electric field is very weak (Friedman, 1963).

In summary, the Hall mobility measures a different effect from that measured by the mobility which enters into the electrical conductivity. Nonetheless the magnitudes of these two mobilities usually become comparable to one another for coherent transport associated with an energy band whose width greatly exceeds the thermal energy κT. Consideration of the complementary limit, coherent transport associated with an energy band that is narrower than the thermal energy, reveals cases in which the two mobilities differ from one another in magnitude and temperature dependence. The sign of the Hall mobility can even be anomalous. Large polarons' motion is generally described as coherent. However, coherent small-polaron motion is only envisioned in extreme circumstances: low temperatures in nearly perfect crystals under the influence of weak applied fields. Otherwise small-polaron motion is described as being incoherent. The Hall effect for such hopping transport is the subject of the remainder of this chapter.

13.2 Microscopic treatment of electric and magnetic fields

Application of electric and magnetic fields alters a charge carrier's inter-site motion. In particular, the spatially varying electric and vector potentials associated with applied fields lift the translational degeneracy of a lattice's equivalent sites. Thus an applied electric field alters a carrier's potential energy as it moves between sites. As described below, a magnetic field's vector potential modifies a carrier's inter-site transfer integral.

For simplicity consider the tight-binding treatment of an excess electron in a crystal with atomic displacements being ignored. In the presence of a spatially constant magnetic field the local electronic function for an excess electron of energy E_g centered on lattice site g can be written as

$$\phi_g(r) = \exp\left[-i\frac{e}{2\hbar}(H \times g)\cdot r\right]\phi(r-g). \tag{13.4}$$

Here $q = -e$, where e is the electronic charge's magnitude and $\phi(r-g)$ maintains the lattice periodicity by satisfying

$$\left\{\frac{1}{2m}\left[\hat{p} + \frac{e}{2}H\times(r-g)\right]^2 + V(r-g)\right\}\phi(r-g) = E_g\phi(r-g) \tag{13.5}$$

when the vector potential is written in its symmetrical gauge $A(r) = (H \times r)/2$. Thus, with a spatially constant magnetic field a carrier's local electronic functions, the $\phi_g(r)$'s, are distinguished from one another by their gauge-dependent phases.

The transfer integral linking local electronic functions at sites g and g' depends on the difference of their two magnetic-field-dependent phase factors: $\exp[-(ie/2\hbar)$ $(g - g') \times r\cdot H)]$, where vector algebra's cyclic rule has been applied. This transfer integral becomes proportional to $\exp[-(ie/4\hbar)(g - g') \times (g + g')\cdot H]$ when its integrand peaks near $r = (g+g')/2$. This factor simplifies to $\exp[-(ie/2\hbar)(g \times g')\cdot H]$ after performing some vector algebra. Thus the energy associated with a carrier's transfer from site g to g' becomes $\exp(i\alpha_{g,g'})t_{g,g'}$, where $\alpha_{g,g'} \equiv -(e/2\hbar)(g \times g')\cdot H$ is termed the magnetic phase factor. When considering the Hall effect, which is first-order in the magnetic field strength, $t_{g,g'}$ is usually taken as just the field-free electronic transfer energy. That is, corrections to $t_{g,g'}$ that are first order in the magnetic field are usually ignorable. With sufficient symmetry these corrections even vanish (Friedman and Holstein, 1963).

Incorporating applied electric and magnetic fields F and H into the differential equations governing a carrier's set of site-occupation amplitudes, Eq. (11.21), the a_g's, yields:

$$i\hbar\frac{\partial a_g}{\partial \tau} = (E_g + eF\cdot g)a_g + \sum_{g'}(t_{g,g'}e^{i\alpha_{g,g'}})a_{g'}. \tag{13.6}$$

This equation incorporates a particular choice of the zero of the electric potential and of the gauge. However, physically observed quantities calculated from these equations are independent of these selections.

Equation (13.6) provides the basis for describing the scattering-type transport addressed in Section 13.1. In particular, a carrier is represented by a wavepacket which evolves under the influence of electric and magnetic fields. Such a wavepacket is constructed from a superposition of Bloch states which are each expressed as a superposition of localized states. The evolution of the localized states' occupation amplitudes are governed by equations of motion, Eq. (13.6). This procedure regains the terms of the Boltzmann equation which describe carriers being driven by an electric field and deflected by a magnetic field via the Lorentz force (Friedman, 1963). The full Boltzmann equation is obtained upon adding its standard relaxation term. Thus the microscopic treatment of field-driven transport of Eq. (13.6) leads to Eq. (13.1), the solution of the Boltzmann equation, for scattering-type transport's Hall mobility.

13.3 Hall mobility for non-adiabatic polaron hopping

To address polaron motion in the presence of electric and magnetic fields the electronic equations of motion Eq. (13.6) must be augmented to incorporate atoms' movements. In particular, the Hamiltonian describing atoms' harmonic displacements is added to the electronic energy in Eq. (13.6), $E_g \rightarrow E_g + H_{\text{vib}}$:

$$ i\hbar \frac{\partial a_g}{\partial \tau} = \left(E_g + H_{\text{vib}} + e\mathbf{F} \cdot \mathbf{g}\right) a_g + \sum_{g'} \left(t_{g,g'} e^{i\alpha_{g,g'}}\right) a_{g'}. \tag{13.7} $$

The electron–phonon interaction manifests itself in the dependence of the local electronic energy E_g and the electronic transfer energies $t_{g,g'}$ on atomic displacements. The microscopic equations of motion Eq. (13.7) provide the basis for discussing the Hall effect for polaron hopping.

The Hall effect for hopping polarons manifests itself through jump rates' dependences on the strengths of the applied electric and magnetic fields. These jump rates are calculated from the equations of motion Eq. (13.7). Jump rates have been computed in two limits. In the adiabatic limit the electronic transfer energies are assumed large enough for an electronic carrier to adjust to atoms' movements. A carrier will then always negotiate a hop in response to appropriate atomic motions. In the non-adiabatic limit the electronic transfer energies are assumed to be so small that an electronic carrier can only rarely negotiate a hop in response to appropriate atomic motions. Electronic transfer energies are often large enough (typically $> \hbar\omega$ and κT) to justify employing the adiabatic limit. However, calculations are easiest in the non-adiabatic limit. Then Eq. (13.7) is solved by treating electronic transfer energies as small perturbations which enable the excess electron to jump between sites.

The simplest geometry for which to address the Hall mobility for polaron hopping is the triangular lattice depicted in panel *a* of Fig. 13.2. This geometry was first used to address the Hall effect for non-adiabatic polaron hopping (Friedman and Holstein, 1963). To lowest order in the transfer energies the non-adiabatic jump rate is proportional to the square of the absolute value of transfer energy connecting the hop's two sites, $|t_{g,g'} \exp(i\alpha_{g,g'})|^2$. Since the magnetic phase factors do not contribute to the absolute value, this term does not contribute to the jump rate's magnetic-field dependence. However, processes that are of higher order in the electronic transfer energies contribute to the magnetic-field dependence of the jump rate. In particular, the lowest-order magnetic-field dependent contributions to the jump rate in the triangular lattice are third order in the electronic transfer energies. These processes, illustrated in panel *a* of Fig. 13.3, involve the interference between the amplitude for a carrier's direct hop between adjacent sites ($\boldsymbol{g} \to \boldsymbol{g'}$) and the amplitude for the carrier's passing through an intermediate site ($\boldsymbol{g} \to \boldsymbol{g''} \to \boldsymbol{g'}$). The magnetic-field dependent portion of the jump rate is proportional to the sum of the magnetic phase factors for the three involved jumps: $\alpha_{g,g''} + \alpha_{g'',g'} - \alpha_{g,g'}$. Utilizing the definition of the magnetic phase factors from above Eq. (13.6) and elementary rules of vector algebra, the sum of the magnetic phase factors is found to equal $e\boldsymbol{H}\cdot\boldsymbol{A}_\Delta/\hbar$, where $\boldsymbol{A}_\Delta \equiv (\boldsymbol{g''} - \boldsymbol{g}) \times (\boldsymbol{g'} - \boldsymbol{g})/2$ is the area of the triangle subtended by the three involved sites. Thus, the magnetic-field-dependent portion of the non-adiabatic jump rate is proportional to the magnetic flux passing through the triangle defined by sites $\boldsymbol{g}$, $\boldsymbol{g'}$ and $\boldsymbol{g''}$.

Studies of the Hall effect for polaron hopping focused on the high-temperature regime $\kappa T > \hbar\omega$ in which atomic motion becomes classical. Atoms' movements

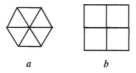

a   *b*

Fig. 13.2 Portions of a triangular lattice and a square lattice are illustrated in panels *a* and *b*, respectively. Hops occur between sites located at these structures' vertices.

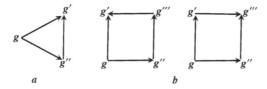

a   *b*

Fig. 13.3 The three-site and four-site interference processes which generate the predominant magnetic-field dependences for non-adiabatic nearest-neighbor polaron hopping in triangular and square lattices are illustrated in panels *a* and *b*, respectively.

are then treated as explicit functions of time. In this regime contributions to the polaron jump rate are thermally activated. The activation energy for the leading field-free contribution to the jump rate is the minimum energy required to displace atoms from the polaron's equilibrium configuration to bring the electronic energies of a hop's initial and final sites into coincidence with one another E_2 (cf. Section 11.5). This two-site activation energy is that for the standard mobility $E_2 = E_A$. The activation energy for the leading magnetic-field-dependent contribution to the jump rate is the minimum energy needed to bring the electronic energies of its three sites into coincidence E_3. Since the constraint for a three-site coincidence exceeds that for a two-site coincidence $E_3 > E_2$. As described in the second paragraph of this chapter, the Hall mobility is the magnetic-field-induced transverse component of a carrier's velocity divided by the product of its longitudinal component and the magnetic field. Thus, the Hall mobility's activation energy is $E_H \equiv E_3 - E_2$. For the short-range electron–phonon interaction envisioned in Holstein's molecular-crystal model $E_n = (F^2/2k)[(n - 1)/n]$ and $E_H \equiv (E_3 - E_2) = E_2/3$.

The lowest-order magnetic-field-independent (second-order) and magnetic-field-dependent (third-order) contributions to the non-adiabatic jump rate for a triangular lattice are (Friedman and Holstein, 1963):

$$R_2 = \left[\frac{t^2}{\hbar} \left(\frac{\pi}{4\kappa T E_2} \right)^{1/2} \right] e^{-E_2/\kappa T} \tag{13.8}$$

and

$$R_3 = \left(\frac{e\mathbf{H} \cdot \mathbf{A}_\Delta}{\hbar} \right) \left[\frac{2}{\sqrt{3}} \frac{t^3}{\hbar} \left(\frac{\pi}{4\kappa T E_2} \right) \right] e^{-E_3/\kappa T}, \tag{13.9}$$

where the field-free geometrically equivalent nearest-neighbor transfer energies are written as $t_{g,g'} \equiv -t$. As described in Section 11.5, for non-adiabatic hopping (1) the duration of a coincidence τ_c is much less than a vibrational period $\omega\tau_c \ll 1$ and (2) the time needed for a carrier's transfer $\tau_t \equiv \hbar/|t|$ is much longer than the time window provided by a coincidence event $\tau_t \gg \tau_c$. Expressed in terms of these times $R_2 \approx v(\tau_c/\tau_t)^2 \exp(-E_2/\kappa T)$ and $R_3 \approx v(e\mathbf{H} \cdot \mathbf{A}_\Delta/\hbar)(\tau_c/\tau_t)^3 (\omega\tau_c)\exp(-E_2/3\kappa T)$, where $v \equiv \omega/2\pi$. The factor $\omega\tau_c$ in R_3 arises from the requirement that atomic vibrations bring the electronic energies of all three sites into coincidence within the time interval τ_c.

The second-order jump rate is simply related to the in-plane mobility for the triangular lattice:

$$\mu_t = -\frac{3}{2}\frac{ea^2}{\kappa T}R_2 = -\left(\frac{ea^2}{\hbar}\right)\left[\frac{3t^2}{2\kappa T}\left(\frac{\pi}{4\kappa TE_2}\right)^{1/2}\right]e^{-E_2/\kappa T}. \tag{13.10}$$

As described in the second paragraph of Section 13.1, the Hall mobility measures the deflection by the magnetic field of carriers drifting in an electric field. As a result, the Hall mobility of a carrier hopping in a triangular lattice depends on the ratio R_3/R_2 (Friedman and Holstein, 1963):

$$\mu_{H,t} = \frac{2}{\sqrt{3}}\frac{R_3}{H_\perp R_2} = \left(\frac{ea^2}{\hbar}\right)\left[\frac{t}{\sqrt{3}}\left(\frac{\pi}{4\kappa TE_2}\right)^{1/2}\right]e^{-(E_3-E_2)/\kappa T}, \tag{13.11}$$

with $E_3 - E_2 = E_2/3$ for the molecular-crystal model. In obtaining Eq. (13.11) the magnetic flux through the triangle $\boldsymbol{H}\cdot\boldsymbol{A_\Delta}$ was written $H_\perp A_\Delta$, where $H_\perp$ is the component of the magnetic field perpendicular to the current flow and the area of the triangular lattice's equilateral triangle is expressed in terms of the inter-site separation a as $A_\Delta = \sqrt{3}a^2/4$. The Hall coefficient is then

$$R_{H,t} = -\frac{\mu_{H,t}}{\sigma} = \frac{1}{ne}\left(\frac{\mu_{H,t}}{\mu_t}\right) = \frac{-1}{ne}\left[\left(\frac{2}{3\sqrt{3}}\right)\frac{\kappa T}{t}\right]e^{(2E_2-E_3)/\kappa T}, \tag{13.12}$$

where $2E_2 - E_3 = 2E_2/3$ for Holstein's molecular-crystal model (Friedman and Holstein, 1963). The Hall coefficient for non-adiabatic polaron hopping in the triangular lattice is seen to depart significantly in magnitude and temperature dependence from its free-carrier value, $-1/ne$.

The magnitudes of the non-adiabatic mobility and Hall mobility for the triangular lattice are both plotted against reciprocal temperature in Fig. 13.4. These plots indicate that the Hall mobility is generally larger than the standard mobility. The temperature dependence of the Hall mobility is also much weaker than that of the standard mobility.

The Hall mobility has also been calculated for non-adiabatic semiclassical polaron hopping between sites of the square lattice depicted in panel b of Fig. 13.2 (Emin, 1971c). The Hall effect for non-adiabatic hopping results from interferences between the transition amplitudes associated with different hopping routes that connect the same initial and final sites. For nearest-neighbor hopping on a square lattice these interference processes require at least four transfers. The corresponding magnetic-field-dependent jump rates are therefore of fourth order in the electronic-transfer energies. Both types of interference process are also proportional to the sums of the process' magnetic phase factors: $\alpha_{g,g''} + \alpha_{g'',g'''} + \alpha_{g''',g'} - \alpha_{g,g'}$ and $\alpha_{g,g''} + \alpha_{g'',g'''} - \alpha_{g',g'''} - \alpha_{g',g'''}$, respectively. Both of these sums of magnetic phase factors equal $e\boldsymbol{H}\cdot\boldsymbol{A_\square}/\hbar$,

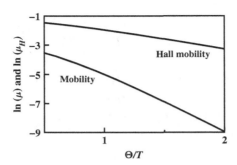

Fig. 13.4 The magnitudes of the non-adiabatic mobility and non-adiabatic Hall mobility in units of $ea^2/\hbar$ for a triangular lattice are plotted versus reciprocal temperature in units of the phonon temperature $\Theta \equiv \hbar\omega/\kappa$. These plots' parameters are $E_2 = 5\hbar\omega$ and $t = \hbar\omega/2$.

where $A_\square \equiv (\mathbf{g}''' - \mathbf{g}) \times (\mathbf{g}' - \mathbf{g}'')/2$ is the area of the square enclosed by the four involved sites, a^2 for inter-site separation a. Thus, the magnetic-field-dependent portions of the non-adiabatic jump rates for a square lattice are proportional to the magnetic flux passing through one of the lattice's elementary units.

As in the case of the triangular lattice, the magnetic-field-dependent contribution to the non-adiabatic semiclassical jump rate for a square lattice is associated with the formation of a triple coincidence. In these instances the electronic energies of initial and final sites of the interference process share a transitory equality with one of the two other sites. The fourth-order magnetic-field-dependent contribution to jump rates in the square lattice is similar to that for the triangular lattice, Eq. (13.9):

$$R_4 = \left(\frac{eH_\perp a^2}{\hbar}\right)\left[\frac{2}{\sqrt{3}}\frac{t^3}{\hbar}\left(\frac{\pi}{4\kappa TE_2}\right)I_4\right]e^{-E_3/\kappa T}, \qquad (13.13)$$

where $H_\perp$ is the component of the magnetic field perpendicular to the plane of the square lattice and the effect of the non-coincident site is contained in the factor I_4. The Hall mobility for non-adiabatic nearest-neighbor hopping in a square lattice also resembles the analogous expression for a triangular lattice, Eq. (13.11):

$$\mu_{H,s} = \frac{4R_4}{H_\perp R_2} = \left(\frac{ea^2}{\hbar}\right)\left[\frac{t}{\sqrt{3}}\left(\frac{\pi}{4\kappa TE_2}\right)^{1/2}(8I_4)\right]e^{-(E_3-E_2)/\kappa T}. \qquad (13.14)$$

The Hall mobility of a square lattice is distinguished from that of a triangular lattice by the presence of the non-coincident site. As illustrated in Fig. 13.5, virtual occupations of non-coincident sites provide bridges which enable transfer among

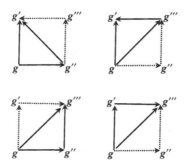

Fig. 13.5 The magnetic-field-dependent contribution to the semiclassical jump rate between sites requires forming a triple coincidence involving the electronic states at these two sites plus that associated with one of the intermediate sites. Virtual transfer through non-coincident sites, depicted with dotted arrows, enable effective transfers, represented by solid arrows, among the three coincident sites. The four pertinent situations relevant to four-site hopping in a square lattice are illustrated above.

trios of coincident sites. The sign and magnitude of a virtual transfer through a non-coincident site depend on the difference between its electronic energy and the electronic energy of the three coincident sites. This difference of electronic energies depends on atomic displacements. In particular, the sign of the virtual transfer changes when alterations of atomic displacements cause the electronic energy of the non-coincident site to overtake the electronic energy of the three coincident sites. This sign change occurs when the four sites experience a quadruple coincidence.

The minimum atomic-displacement energy needed to establish a quadruple-coincidence is denoted E_4. The energy characterizing a deformation-related change of the sign of the virtual transfer involved in the semiclassical Hall effect is $E_4 - E_3$. This energy difference is generally a small fraction of E_2. For example, $E_4 - E_3 = E_2/6$ for Holstein's molecular-crystal model. The mobility activation energy E_2 is frequently estimated from experiment. In many instances $E_4 - E_3$ is less than the thermal energy κT where mobility measurements are performed.

The introduction of the factor I_4 in Eq. (13.13) and Eq. (13.14) [as compared with Eq. (13.9) and Eq. (13.11)] results from the elementary structural unit of the square lattice having one more site than that of the triangular lattice. In particular, the virtual transfer energy averaged over thermally induced atomic displacements is essentially tI_4 (Emin, 1971c). For classical atomic motion $I_4 \rightarrow t/E_3$ when $E_4 - E_3 \gg \kappa T$. Then $\mu_{H,s}$ only differs from $\mu_{H,t}$ by a temperature-independent factor. By contrast, $I_4 \rightarrow (t/4\kappa T)\exp[-(E_4 - E_3)/\kappa T]$ when $E_4 - E_3 < \kappa T$. In these instances $\mu_{H,s}$ manifests a different temperature dependence than $\mu_{H,t}$. This feature is evident in Fig. 13.6 where $\mu_{H,t}$ and $\mu_{H,s}$ are both plotted against reciprocal temperature.

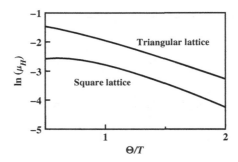

Fig. 13.6 The non-adiabatic Hall mobilities in units of $ea^2/\hbar$ for triangular and square lattices are plotted versus reciprocal temperature in units of the phonon temperature $\Theta \equiv \hbar\omega/\kappa$. These plots' parameters are $E_2 = 5\hbar\omega$ and $t = \hbar\omega/2$.

The Hall coefficient for nearest-neighbor hopping on a square lattice becomes:

$$R_{H,s} = -\frac{\mu_H}{\sigma} = \frac{1}{ne}\frac{\mu_{H,s}}{\mu_s} = \frac{-1}{ne}\left[\frac{\kappa T}{t}\frac{8I_4}{\sqrt{3}}\right]e^{(2E_2-E_3)/\kappa T} \tag{13.15}$$

upon recalling the Hall mobility Eq. (13.14) and noting that a square lattice's mobility is

$$\mu_s = -\frac{ea^2}{\kappa T}R_2 = -\left(\frac{ea^2}{\hbar}\right)\left[\frac{t^2}{\kappa T}\left(\frac{\pi}{4\kappa T E_2}\right)^{1/2}\right]e^{-E_2/\kappa T}. \tag{13.16}$$

When $\kappa T > E_4 - E_3$ the Hall coefficient rises exponentially with increasing reciprocal temperature, $R_{H,s} \sim (-1/ne)\exp[(2E_2 - E_4)/\kappa T]$ with $2E_2 > E_4$. Similar behavior persists even if the electron–phonon coupling is so strong that $E_4 - E_3 \gg \kappa T$: $R_{H,s} \sim [(-1/ne)(5\kappa T/E_3)\exp[(2E_2 - E_3)/\kappa T]$ with $2E_2 > E_3$. Thus the mobility ratio produces a temperature dependence which mimics that from a semiconductor's thermally activated carrier density.

All told, the non-adiabatic limit of semiclassical polaron hopping produces a distinctive Hall effect. The Hall effect arises as atoms' vibrations cause the electronic energies of a hop's initial and final sites to come into momentary coincidence with the electronic energy of another site. The magnitude of the resulting Hall mobility is generally larger than that of the conventional mobility. In addition, the Hall mobility's temperature dependence is markedly weaker than the thermally activated behavior of the standard hopping mobility. The large disparity between the temperature dependences of the Hall mobility and the conventional mobility leads to a Hall coefficient whose temperature dependence often resembles that arising from a thermally activated carrier density. These distinguishing features can be utilized to identify polaron hopping.

13.4 Hall mobility for adiabatic polaron hopping

Hopping is non-adiabatic when the governing electronic transfer energy is so small that electronic carriers usually cannot transfer fast enough to negotiate a hop in response to suitable atomic motion. Such electronic transfer energies are typically less than the characteristic phonon energy, $t < \hbar\omega$. By contrast, adiabatic hopping occurs when electronic transfer energies are large enough to enable carriers to follow atomic movements which are themselves modified by the electronic carriers. The Hall effect has been calculated for small-polaron hopping among sites of Holstein's molecular-crystal model arranged in a triangular lattice (Emin and Holstein, 1969).

An applied magnetic field generally alters a carrier's wavefunction. As described in Section 13.2 this change causes the carrier's electronic transfer energies to garner magnetic-field-dependent phase factors. These magnetic phase factors affect the adiabatic Hamiltonian governing atomic motion. In particular, these magnetic phase factors generate those terms in the adiabatic Hamiltonian of Eq. (4.5) which are proportional to the momenta P that are conjugate to atomic displacement parameters associated with the reduced mass M. As shown in Eq. (4.5) these terms have the form $e(A \cdot P + P \cdot A)/2M$, where Eq. (4.6) defines the fictitious vector potential A. This fictitious vector potential is proportional to the actual applied magnetic field H.

Hopping transport envisions jumps of a carrier between discrete sites. Therefore a carrier's location is indicated by the centroid of the occupied site g rather than by the continuous variable r. Concomitantly, the carrier's wavefunction is designated by the set of occupation amplitudes $\{c_g\}$ rather than by the continuous function $\phi(r)$ used in Eq. (4.6). Furthermore, the strain $\Delta(u)$ at position u which characterizes the deformable continuum described in Chapter 4 is replaced by a set of local deformation parameters $\{x_g\}$. For Holstein's molecular-crystal model the energy of a carrier at site g is presumed to depend linearly on the deformation parameter which is assigned to the same site: $E_g \equiv -Fx_g$. The scalar F then denotes the magnitude of this short-range electron–phonon interaction. Using Eq. (13.6) the electronic amplitudes corresponding to the electronic eigenvalue $W(\ldots x_g \ldots)$ are found to satisfy

$$Wc_g = \left(-Fx_g + eF \cdot g\right)c_g + \sum_{g'}\left(t_{g,g'}e^{i\alpha_{g,g'}}\right)c_{g'}. \tag{13.17}$$

A carrier executes a thermally activated semiclassical hop from site g to site g' when atomic deformation parameters change in time so that $|c_g|^2$ falls from near unity to near zero as $|c_{g'}|^2$ rises from near zero to near unity. These changes of sites' occupation probabilities primarily occur when the electronic energies of the two sites are within $|t_{g,g'}| \equiv t$ of being coincident with one another. The transfer energy t is generally smaller than the deformation energy associated with establishing a coincidence $F^2/4k$ since small-polaron formation requires that $F^2/2k \gg t$. Alternatively stated, the conformational change characterizing a coincidence's establishment $F/4k$ exceeds that characterizing an electronic carrier's transfer t/F. In addition, the deformation energy associated with forming a coincidence is usually much larger than the thermal energy: $F^2/4k \gg \kappa T$. In other words, a coincidence's characteristic conformal change is much larger than atomic vibrations' thermal amplitudes: $F/4k \gg (\kappa T/k)^{1/2}$, where k is the deformation parameter's stiffness constant. Thus, atomic vibrations only sporadically enable a small polaron to execute an adiabatic hop.

The Hall effect for nearest-neighbor hopping in a triangular lattice can be addressed by considering hopping among three mutually adjacent sites. A Hall effect results when the magnetic field alters atomic motion so as to change the final site of a hop. Then, for example, a carrier which would have hopped from site 1 to site 2 in the absence of the magnetic field jumps to site 3 when the magnetic field is applied.

The relative values of the three sites' atomic displacement parameters determine which site the carrier occupies. An orthogonal transformation of the molecular-crystal-model's deformation parameters for the three sites, x_1, x_2 and x_3, yields two relative displacement coordinates, x and y:

$$x \equiv \frac{x_2 + x_3 - 2x_1}{\sqrt{6}} \tag{13.18}$$

and

$$y \equiv \frac{x_3 - x_2}{\sqrt{2}}, \tag{13.19}$$

with $z \equiv (x_1 + x_2 + x_3)/\sqrt{3}$. Since c_1, c_2 and c_3 only depend on x and y, semiclassical hops are described by these variables' evolution with time τ. The classical trajectories, $x(\tau)$ and $y(\tau)$, are governed by the adiabatic Hamiltonian. The adiabatic Hamiltonian for the three pertinent sites contains a vibratory contribution $k(x^2 + y^2 + z^2)/2$ and an electronic contribution $W = \varepsilon(x, y) - Fz/\sqrt{3}$.

Three electronic eigenstates are found by solving Eq. (13.17) for three mutually adjacent sites. Upon taking the nearest-neighbor transfer energies to be negative, $t_{g,g'} = -t$ with $t > 0$, the lowest energy electronic eigenstate is found to be nodeless and non-degenerate. The corresponding energy $\varepsilon(x,y)$ then comprises a triangular pyramid whose edges become rounded for finite t. This rounding of the pyramid's apex

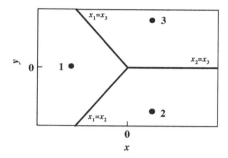

Fig. 13.7 The space defined by the configuration coordinates x and y consists of regions labeled as 1, 2, and 3. Each region's label indicates the site occupied by the carrier for configuration coordinates located in that region. The boundaries between the three regions are defined by coincidence conditions between the corresponding sites' deformation parameters. The minimum of the adiabatic potential within each region is indicated with a large dot.

reduces its energy from $\varepsilon(0,0) = 0$ to $\varepsilon(0,0) = -2t$. As illustrated in Fig. 13.7 the three edges of the triangular pyramid coincide with the coincidence lines $x_1 = x_2 (y = x\sqrt{3}$ for $x \leq 0)$, $x_1 = x_3 (y = -x\sqrt{3}$ for $x \leq 0)$, and $x_2 = x_3 (y = 0$ for $x \geq 0)$. These three coincidence lines divide the xy configuration space into three equivalent regions within which the occupation probabilities, the $|c_g|^2$, for each of the three corresponding sites approach unity. Each region is therefore labeled by the corresponding occupied site.

In the limit $t \to 0$ the rounding of the pyramid's edges disappears. Then the pyramidal electronic energy contour assumes a simple form:

$$\varepsilon(x,y) = \frac{F}{\sqrt{6}}(2x), \qquad (13.20)$$

$$\varepsilon(x,y) = -\frac{F}{\sqrt{6}}\left(x - \sqrt{3}y\right) \qquad (13.21)$$

and

$$\varepsilon(x,y) = -\frac{F}{\sqrt{6}}\left(x + \sqrt{3}y\right) \qquad (13.22)$$

in regions 1, 2 and 3, respectively. Combining $\varepsilon(x, y)$ with the vibrational potential $k(x^2 + y^2)/2$ yields a potential well with minima of $-F^2/3k$ located in each of the three regions. These minima are at $\left(-2F/k\sqrt{6}, 0\right)$, $\left(F/k\sqrt{6}, -F/k\sqrt{2}\right)$ and $\left(F/k\sqrt{6}, F/k\sqrt{2}\right)$ in regions, 1, 2, and 3, respectively. A semiclassical adiabatic hop between sites is represented by a trajectory which passes between the corresponding potential wells.

A reduced adiabatic Hamiltonian governs the atomic motions which dictate the excess electron's movement among three mutually adjacent sites of a triangular molecular-crystal lattice:

$$H_3 = \frac{P_x^2 + P_y^2}{2M} + \frac{k}{2}\left(x^2 + y^2\right) + \varepsilon(x, y) + \frac{e}{2M}\left(\boldsymbol{A}\cdot\boldsymbol{P} + \boldsymbol{P}\cdot\boldsymbol{A}\right). \qquad (13.23)$$

The first three contributions are (1) the kinetic-energy associated with movement within the xy configuration coordinate plane, (2) the potential energy of these configuration coordinates' vibrations and (3) the electron's energy as a function of these configuration coordinates. The remaining term is central to the Hall effect. It couples the momenta conjugate to the x and y configuration coordinates with the fictitious vector potential. The fictitious vector potential results from the magnetic-field dependence of the electronic carrier's inter-site transfer in response to changes of the configuration coordinates. The magnetic field generated from the fictitious vector potential is (1) proportional to the applied magnetic field, (2) directed perpendicular to the xy plane, and (3) strongly peaked about the origin, the triple-coincidence point. The decrease of the fictitious magnetic field's strength with distance from the origin is (1) generally characterized by the configuration coordinate distance $\sim t/F$ and (2) weakest along the three double coincidence lines. The total flux of the fictitious magnetic field through the xy configuration coordinate plane equals that of the applied magnetic field through the real-space triangle whose vertices are three mutual nearest-neighbor sites (Emin and Holstein, 1969).

Semiclassical adiabatic hops between sites are depicted by trajectories which pass between potential wells in the xy configuration space. Thus an adiabatic hop from site 1 to site 3 corresponds to a trajectory which passes from the potential well of region 1 to that of region 3. These excursions are governed by the Hamiltonian of Eq. (13.23). Vibrational dispersion is not explicitly present in this reduced three-site Hamiltonian. Nonetheless its implicit presence enables the fictitious particle to exchange energy with the remainder of the lattice before and after a hop. As discussed in Section 11.6, this energy transfer is presumed sufficient for each inter-well passage to be regarded as an independent event. Thus the fictitious particle whose trajectory describes a thermally assisted hop usually garners energy before leaving its initial region and deposits energy after entering its final region.

The adiabatic jump rate associated with a pair of sites is the probability per unit time of a trajectory passing between the corresponding regions. To count the relevant trajectories it is useful to construct a reference line deep within the region

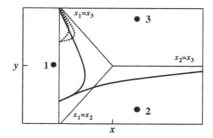

Fig. 13.8 At low energies the trajectories from a position on the reference line which can reach region 3 are bracketed by two trajectories (dotted lines) which are tangent to the $x_1 = x_3$ coincidence line. At high energies the trajectories from a position on the reference line which can reach region 3 are bracketed by one solid-line trajectory which is tangent to the $x_1 = x_3$ coincidence line and another solid-line trajectory which is tangent to the $x_2 = x_3$ coincidence line.

corresponding to the trajectories' initial site. The reference line for trajectories leaving region 1 is the line $x = -d$, where $2F/k\sqrt{6} > d \gg t/F$. A trajectory leaving region 1 can be characterized by its energy E and by the transverse component of velocity v_y it possesses as it crosses the reference line. Then, as illustrated in Fig. 13.8, trajectories of energy E crossing the reference line of region 1 at $y = y_0$ will reach region 3 for values of v_y between v_y^{max} and v_y^{min}. If E is less than that required for trajectories to reach the peak of the adiabatic potential at $x = y = 0$ then both of these limiting trajectories will be tangential to the $x_1 = x_3$ coincidence line. However, if E is sufficient for trajectories to reach the peak of the adiabatic potential then one of the limiting trajectories for passing from region 1 to region 3 will be tangential to the $x_1 = x_3$ coincidence line while the other is tangential to the $x_2 = x_3$ coincidence line.

The application of a magnetic field shifts the limiting trajectories from each point on the reference line (Emin and Holstein, 1969). In particular, an arbitrarily weak magnetic field just alters the y-components of the fictitious-particle's velocities for the two limiting trajectories which cross the reference line at y_0 by Δv_y^{max} and Δv_y^{min}. The net change of the probability per unit time and per unit energy passing from region 1 to region 3 induced by the magnetic field is then

$$W_{13}^H(E) = \frac{e^{-E/\kappa T}}{MZ} \int_{y_1}^{y_2} dy_0 \left(\Delta v_y^{max} - \Delta v_y^{min} \right). \qquad (13.24)$$

Here $y_2(E)$ and $y_1(E)$ are the limits of y_0 between which trajectories of energy E reach region 3. The partition function for a mass M vibrating with frequency $v \equiv (k/M)^{1/2}/2\pi$ within the quasi-harmonic potential well of depth $-F^2/3k = -E_3$ located in region 1 is

$$Z \approx \left(\frac{\kappa T}{Mv}\right)^2 \exp(E_3/\kappa T).$$ (13.25)

Here E_3 is recognized as the three-site coincidence energy for Holstein's molecular-crystal model defined in the paragraph before Eq. (13.8).

A theorem enables the magnetic-field-dependent contribution to the jump probability from trajectories of energy E, Eq. (13.24), to be expressed in terms of the flux of the fictitious magnetic field enclosed by limiting trajectories of energy E (Emin and Holstein, 1969). In particular,

$$\int_a^b dy_0 \Delta v_y = -\oint_C d\boldsymbol{l} \cdot \Delta \boldsymbol{v} = \frac{-e}{M} \Phi_C(E),$$ (13.26)

where $\Phi_C(E)$ is the flux of the fictitious magnetic field enclosed by the contour of integration C. The integration contour follows (1) the limiting trajectory from $y_0 = a$ on the reference line to asymptotically approach a coincidence line, (2) the limiting asymptotic trajectory from this coincidence line to the reference line at $y_0 = b$ and (3) the reference line from $y_0 = b$ to $y_0 = a$.

This theorem enables Eq. (13.24) to be expressed in terms of $\Phi(E)$, the flux enclosed by limiting trajectories (Emin and Holstein, 1969):

$$W_{13}^H(E) = e\left(\frac{v}{\kappa T}\right)^2 e^{-(E_3+E)/\kappa T} \Phi(E).$$ (13.27)

When $E < -2t$, the energy of the triplet-coincidence peak, trajectories lack sufficient energy to reach the $x_2 = x_3$ coincidence line from the region 1 reference line. Then $\Phi(E) = 0$ as all limiting trajectories are tangent to the $x_1 = x_3$ coincidence line. By contrast, for $E \geq -2t$ some trajectories can reach the $x_1 = x_3$ coincidence line. As illustrated in Fig. 13.9 $\Phi(E)$ is then the flux enclosed between the three equivalent limiting trajectories which are each tangent to two coincidence lines. The enclosed flux increases with energy as the limiting trajectories move further from the triple-coincidence point to encompass an increasing area of the xy plane.

Integrating $W_{13}^H(E)$ over energy yields the magnetic-field-dependent contribution to the adiabatic jump rate in a triangular lattice, analogous to R_3 of Eq. (13.9) for non-adiabatic hopping:

$$R_3^A = (e\boldsymbol{H} \cdot \boldsymbol{A_\Delta})\left(\frac{v^2}{\kappa T}\right) e^{-(E_3-2t)/\kappa T} F(T),$$ (13.28)

where

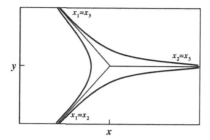

Fig. 13.9 Three equivalent limiting trajectories that are each tangent to two coincidence lines define the border which encloses the magnetic flux $\Phi(E)$. These trajectories move further from the coincidence lines and from the triple coincidence point as their energy E increases. The effective magnetic field's strength is greatest along the coincidence lines and has its absolute maximum at the triple coincidence point. Therefore the enclosed magnetic flux rises as these limiting trajectories' energy is increased.

$$F(T) \equiv \int_{-2t}^{\infty} \frac{dE}{\kappa T} e^{-(E+2t)/\kappa T} \frac{\Phi(E)}{H \cdot A_\Delta}. \tag{13.29}$$

This function increases monotonically from zero to unity as $\kappa T/t$ is raised from well below 1 to well above 1 (Emin and Holstein, 1969). The component of the adiabatic jump rate which is independent of the magnetic field is just

$$R_2^A = v e^{-(E_2-t)/\kappa T}. \tag{13.30}$$

These results are utilized to obtain the adiabatic Hall and standard mobilities for semiclassical nearest-neighbor polaron hopping in the triangular lattice:

$$\mu_{H,t}^A = \frac{2}{\sqrt{3}} \frac{R_3^A}{H_\perp R_2^A} = \left(\frac{ea^2v}{2\kappa T}\right) F(T) e^{-(E_3-E_2-t)/\kappa T} \tag{13.31}$$

and

$$\mu_t^A = -\frac{3}{2} \frac{ea^2v}{\kappa T} e^{-(E_2-t)/\kappa T}, \tag{13.32}$$

where it is recalled that $A_\Delta = \sqrt{3}a^2/4$ for the inter-site separation a. Finally, these results are combined to yield the Hall coefficient for adiabatic semiclassical polaron hopping in the triangular lattice:

$$R_{H,t}^A = -\frac{\mu_{H,t}}{\sigma} = \frac{1}{ne}\left(\frac{\mu_{H,t}^A}{\mu_t^A}\right) = \frac{-1}{ne}\left[\frac{F(T)}{3}\right] e^{(2E_2-E_3)/\kappa T}. \tag{13.33}$$

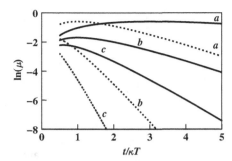

Fig. 13.10 The magnitudes of the triangular lattice's Hall mobilities (solid curves) and standard mobilities (dotted curves) are plotted in units of ea^2v/t against $t/\kappa T$. The three pairs of curves labeled as a, b and c denote progressively increasing values of the ratio of the E_2/t.

For the molecular-crystal model simple relationships exist among the deformation-related contributions to the characteristic energies of the Hall mobility, standard mobility and Hall constant: $E_3 - E_2 = E_2/3$ and $2E_2 - E_3 = 2E_2/3$.

The Hall-mobility activation energy of Eq. (13.31) is much smaller than the standard-mobility activation energy of Eq. (13.32). As a result the Hall mobility manifests a much weaker rise with increasing temperature than does the standard mobility. As illustrated in Fig. 13.10, the Hall mobility may even fall with increasing temperature. Except when the temperature is so high that the thermal energy over-whelms effects of the activation energies, the Hall mobility is larger than the conventional mobility.

The adiabatic approach applies when electronic transfer energies are large enough to preclude transitions between different adiabatic energy surfaces. By contrast, the non-adiabatic regime pertains when transitions between adiabatic surfaces domi-nate. As described in Section 11.6, the Landau–Zener analysis of the semiclassical transition between these two regimes indicates that it occurs when the transfer energy's magnitude becomes comparable to $t_{\mathrm{m}} \sim (\hbar\omega)^{1/2}(E_2\kappa T)^{1/4}$. As expected, the standard mobilities for semiclassical polaron hopping calculated in the adiabatic and non-adiabatic limits, Eqs. (13.32) and (13.10), become comparable to one another when $t \sim t_{\mathrm{m}}$. However, the ratio of the polaron-hopping Hall mobilities calculated in the semiclassical adiabatic and non-adiabatic limits, Eqs. (13.31) and (13.11), becomes $(t_{\mathrm{m}}/\kappa T)\exp(t_{\mathrm{m}}/\kappa T)F(T)$ when $t \sim t_{\mathrm{m}}$. This ratio becomes much less than unity in the high-temperature limit where atomic motion becomes classical $t_{\mathrm{m}}/\kappa T \ll 1$ (Emin and Holstein, 1969). Thus, the adiabatic Hall mobility is much less than the non-adiabatic Hall mobility in the transition regime $t \sim t_{\mathrm{m}}$. This finding suggests that the Hall mobility does not monotonically rise as the transfer energy increases upon passing from the non-adiabatic to the adiabatic limits. Such non-monotonic behavior implies a complicated transition between the semiclassical Hall mobility's non-adiabatic and adiabatic limits.

The transition from non-adiabatic to adiabatic regimes is more complex for the semiclassical Hall mobility than for the standard mobility. In both regimes the standard mobility arises from trajectories which pass through a two-site coincidence. By contrast, different types of trajectory dominate the triangular lattice's Hall mobility in non-adiabatic and adiabatic regimes. In particular, the non-adiabatic Hall effect is dominated by trajectories which pass near the triple coincidence point (Friedman and Holstein, 1959). However the adiabatic Hall effect is dominated by trajectories which ride along the crest which defines a coincidence between a hop's two possible final sites (Emin and Holstein, 1969). Thus the transition between the semiclassical Hall mobility's non-adiabatic and adiabatic regimes accompanies a change of the weighting of these two different types of trajectory.

Finally, the three adiabatic energy surfaces for the three-site configuration, obtained from the eigenvalues of Eq. (13.17), are not equivalent with respect to inversion. This asymmetry is greatest near the origin. At the origin the energy of the lowest adiabatic surface is $\varepsilon(0,0) = -2t$ while the energies of the two higher-lying surfaces both equal t. Thus the adiabatic energy surface for a single excess electron hopping among otherwise vacant sites is not identical to that for a single unoccupied state hopping among sites that are otherwise occupied by electrons. Therefore the Hall-effect dynamics of a solitary excess electron hopping in a triangular lattice is not equivalent to that of a solitary vacant state. In particular, the degeneracy of two energy surfaces at the origin fosters non-adiabatic transitions between them. In addition, electronic degeneracy complicates the determination of the fictitious magnetic field associated with the higher-lying adiabatic surfaces. The t-dependences of the Hall-mobility activation energy and of the energy characterizing the Hall coefficient are also altered: cf. Eqs. (13.28), (13.31) and (13.33). Finally, the Hall-effect sign, addressed in Section 13.5, is not altered upon changing the carrier hopping in a triangular lattice from being an excess electron to being a vacant state (Holstein, 1973).

13.5 Hall-effect signs for hopping conduction

The Hall-effect sign indicates the nature of the charge carriers in a conventional wide-band semiconductor. Electrons that occupy states of positive effective mass near conduction-band minima produce a Hall coefficient of negative sign. By contrast, "holes", carriers associated with vacant electronic states having negative effective masses near valence-band maxima, produce a Hall coefficient of positive sign.

These simple relationships no longer apply when the widths of carriers' energy bands are less than the thermal energy κT. In particular, a different rule applies to carriers that move by hopping conduction. The sign of the Hall coefficient then depends on (1) the charge of the mobile specie, (2) the filling of the band of states amongst which carriers move, and (3) the signs of the electronic-transfer energies associated with the carriers' Hall effect.

This chapter's prior discussion of the Hall effect for polaron hopping explicitly considered a single excess electron hopping among sites that are empty. The complementary problem addresses the Hall effect for electronic hopping when all but a single site is occupied (Holstein, 1973). Passing from the empty-band limit to the full-band limit changes the sign of the Hall coefficient by the factor $(-1)^{n+1}$, where n denotes number of electronic transfers required to complete the closed loop associated with the Hall-effect interference process. With nearest-neighbor transfers $n = 3$ for the triangular lattice and $n = 4$ for the square lattice. Thus, the Hall coefficient for the triangular lattice does not change sign upon transforming from the hopping of a single excess electron to that of a single electronic vacancy: $(-1)^{n+1} = 1$ for $n = 3$. By contrast, the Hall coefficient for the square lattice changes sign upon transforming from the hopping of a single excess electron to that of a single electronic vacancy: $(-1)^{n+1} = -1$ for $n = 4$. This simple rule establishes the relative Hall-coefficient sign upon transforming from the hopping of an excess electron to that of an electronic vacancy.

Combining this rule with generalizations of the Hall coefficients obtained for excess electrons, described in the preceding sections of this chapter, yields a formula for the absolute sign of the Hall coefficient (Emin, 1977a):

$$\operatorname{sgn}(R_H) = \operatorname{sgn}\left[(-q)(\delta)^{n+1}\prod_{i=1}^{n} t_{i,i+1}\right]. \qquad (13.34)$$

Here q denotes the sign of the bona fide carriers (usually electrons, $q = -e$). The filling of the band of states between which carriers hop is characterized by δ: $\delta = -1$ denotes a nearly empty band within which carriers move while $\delta = 1$ denotes a nearly filled band within which vacancies move. The final factor results from hopping conduction's Hall effect depending on the product of electronic transfer energies $t_{g,g+1}$ around a closed loop of n elements. Evaluation of Eq. (13.34) for simple models suggests explanations of why electronic hopping conduction often yields Hall coefficients with unconventional signs.

Consider electrons hopping ($q = -e$) between the nearly empty ($\delta = -1$) anti-bonding orbitals upon which conduction bands of common covalent semiconductors are based. A three-member closed loop of anti-bonding orbitals is schematically illustrated in panel a of Fig. 13.11. Each anti-bonding orbital is antisymmetric about the centroid of the line connecting the corresponding pair of neighboring atoms. The transfer integral that links adjacent anti-bonding orbitals garners its principal

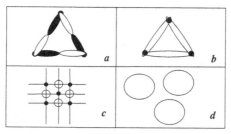

Fig. 13.11 Panel *a* depicts three atoms linked by three anti-bonding orbitals. The antisymmetric character of each of these orbitals is indicated by one of its two lobes being represented in black and the other lobe shown in white. The Hall coefficient for transport via such odd-membered loops is positive, independent of band filling. Panel *b* depicts three atoms linked by three bonding orbitals. The Hall coefficient for transport via such odd-membered loops is negative, independent of band filling. Panel *c* illustrates the structure of MnO. The small solid dots indicate the Mn cations and the large white circles denote the oxygen anions. The solid lines indicate that transfer between the cations is presumably facilitated by the intervening oxygen anion. The Hall coefficient is negative for nearly empty bands and positive for nearly filled bands. Panel *d* illustrates configurations of three mutual neighbors which usually dominate the Hall coefficient of randomly placed localized electronic states.

contribution from the region surrounding their mutually shared atom where these anti-bonding orbitals most strongly overlap. In this region the energy entering into the transfer integral is negative. In this circumstance the product of transfer energies for the closed loop is positive: e.g. $[(-1) \times (-1)]^n > 0$. Upon incorporating these results into Eq. (13.34) the sign of the Hall coefficient is found to be just the sign of δ^{n+1} with $\delta = -1$. Therefore nearest-neighbor hopping of excess electrons between anti-bonding orbitals arranged in odd-member loops (n is odd) generates a positive Hall coefficient.

Consider electrons hopping ($q = -e$) between the nearly filled ($\delta = 1$) bonding orbitals upon which highest-lying valence bands of common covalent semiconductors are based. A three-member closed loop of bonding orbitals is schematically illustrated in panel *b* of Fig. 13.11. With (1) the bonding orbitals being symmetric about the bond's mid-point, and (2) the energy entering into each transfer integral also being negative, the product of transfer energies is $(-1)^n$. Upon incorporating these results into Eq. (13.34) the sign of the Hall coefficient is also found to be just the sign of $(-1)^n$. Therefore nearest-neighbor hopping of an electronic vacancy between bonding orbitals arranged in odd-member loops (n is odd) generates a negative Hall coefficient.

Hall-effect sign anomalies have been reported in amorphous covalent semiconductors for which there is evidence of hopping conduction. Positive Hall coefficients are reported when both amorphous silicon and amorphous arsenic have been doped to introduce electrons into their conduction states (LeComber *et al.*, 1977; Mytilineou and Davis, 1977). In addition, negative Hall coefficients are reported when amorphous germanium and amorphous silicon are doped to remove electrons from their valence states (Clark, 1967; Seager *et al.*, 1974; LeComber *et al.*, 1977). Unlike their crystalline versions, the structures of these amorphous covalent semiconductors are thought to contain significant numbers of odd-member rings (Thorpe *et al.*, 1973; Joannopolous and Cohen, 1973; Robertson, 1975; Paul and Connell, 1976).

Hopping also occurs between localized electronic states that are not involved in covalent bonding. These sites can be those of self-trapped carriers in crystals. Examples include individual cation sites in some transition-metal compounds (e.g. Mn sites in MnO) and sites of V_K-centers (Cl_2^- units) for self-trapped holes in KCl. A portion of the MnO lattice is schematically depicted in panel *c* of Fig. 13.11. Hopping also takes place among non-bonding states in non-crystalline solids. Examples of these include lone-pair states associated with the uppermost valence bands in a number of glasses. In addition, electronic hops proceed between defect states and among impurity states including donor states in lightly doped conventional semiconductors. A non-crystalline arrangement of the orbitals of three mutual neighbors is schematically shown in panel *d* of Fig. 13.11.

Equation (13.34) gives the sign of the Hall coefficient in each of these instances of electronic hopping ($q = -e$) as just the sign of $\delta^{n+1}(-1)^n$, where (as with bonding orbitals) the final factor is the sign of the product of the *n* transfer energies in the pertinent loop. Thus the Hall coefficient is positive for hopping between nearly filled ($\delta = 1$) nearest-neighbor sites of a cubic crystal ($n = 4$). For example, the hopping of electronic vacancies introduced by very sparse substitution of Li for Mn in cubic MnO produces a positive Hall coefficient (Crevecoeur and de Wit, 1970). By contrast, the Hall coefficient is negative for hopping among randomly placed sites whose predominant loops link three mutually adjacent neighbors ($n = 3$). Indeed, hopping in chalcogenide glasses often produces negative Hall coefficients even when the Seebeck coefficients are positive (Emin *et al.*, 1972; Seager *et al.*, 1973, Grant *et al.*, 1974; Nagels *et al.*, 1974; Klaffke and Wood, 1976).

Finally, a situation is noted in which the bona fide carriers are holes ($q = e$). At very low temperatures holes in doped conventional semiconductors are bound in *s*-like shallow acceptor states. At low enough acceptor densities these holes can hop between acceptors. The sign of the Hall coefficient associated with this hopping conduction is then $-\delta^{n+1}(-1)^n$. With three-member loops dominating the Hall effect at low acceptor concentrations, the Hall coefficient is always positive. In particular, the Hall coefficient remains positive even though the Seebeck coefficient will change sign with compensation (e.g. changing δ from -1 to 1). An analogous situation occurs for inter-donor hopping ($q = -e$) in conventional semiconductors. The Hall coefficient then remains negative while the Seebeck coefficient changes sign with compensation (Geballe, 1959).

Part III

Extending Polaron Concepts

14

Superconductivity of large bipolarons

As the temperature is lowered below a critical value the transport properties of some materials display the distinctive features which taken together define "superflow". The superflow of electrons and helium atoms describe superconductivity and superfluidity, respectively. With superconductivity electrons flow without resistance and with a vanishing Seebeck coefficient while expelling magnetic flux (the Meissner effect). These features imply that carriers manifest superconductivity when they condense into a collective ground state which precludes their scattering and their orbital paramagnetism. Analogously, helium atoms manifest superfluidity when they condense into a collective ground state which precludes their scattering and their rotational flow.

The mobile species of these collective ground states are bound together by their mutual attraction. The van der Waals (electrostatic) attraction between helium atoms drives their condensation into a liquid. The superfluid-^{4}He phase is nested within its liquid phase. A phonon-mediated attraction among the conduction electrons of a metal having energies close to its Fermi level produces the condensate for BCS superconductivity (Bardeen *et al.*, 1957a; 1957b). In particular, the attraction between electrons is mediated by their mutual coherence with phonons. The stabilization of the condensate depends on electron's coherent response to atoms' zero-point dynamics.

It has long been conjectured that mobile real-space singlet pairs might form the basis of a superconducting condensate analogous to that for liquid ^{4}He (Schafroth, 1955). Small- and large-singlet bipolarons constitute real-space singlet pairs associated with displacing the equilibrium positions of surrounding atoms. As discussed in Chapter 10, small polarons are generally severely localized and move incoherently by thermally assisted hopping. As such they are not expected to be a suitable basis for bipolaronic superconductivity (Anderson and Abrahams, 1987; Emin and Hillery, 1989; Chakraverty *et al.*, 1998). By contrast, in analogy with large polarons, large bipolarons are presumed to move coherently.

To obtain a form of superconductivity that is akin to the superfluidity of ^{4}He, interactions between large bipolarons would have to facilitate their condensing into a liquid. The ground state of this interacting system would also have to remain fluid rather than solidifying. Finally, the excitation spectrum of this fluid ground state would then have to support resistance-less flow and the Meissner effect. A discussion of this scenario begins by addressing the interactions between large bipolarons.

14.1 Interactions between large bipolarons

Large bipolarons, each carrying the electronic charge of two carriers, repel one another via their long-range Coulomb repulsion. These repulsions are modulated by the static dielectric constant ε_0 since self-trapped carriers can move no more rapidly than the atoms whose displacements bind them. Thus, the Coulomb repulsion between two bipolarons of electronic charge $2e$ separated by the distance s is written as $(2e)^2/\varepsilon_0 s$, where s exceeds a bipolaron's radius R_b (Emin, 1994a; 1994b).

An additional repulsive effect comes into play as two large bipolarons get close enough to one another to begin to merge into a "quadpolaron". In particular, as illustrated in Fig. 7.1, the Pauli principle dictates that only one singlet pair occupies the lowest non-degenerate level of the self-trapping potential well. The energy associated with promoting two carriers above the self-trapping well's lowest level destabilizes a large quadpolaron with respect to two separate large bipolarons (Emin, 1994b). In other words, large bipolarons experience a short-range repulsion of one another. This repulsion has been modeled as a hard-core repulsion akin to that between ^{4}He atoms: $\varepsilon_S(s) = C_1 \exp(-C_2 s/R_b)$, where s is the separation between two large bipolarons having radii R_b while C_1 and C_2 denote numerical constants (Emin, 1994a; 1994b).

Attractive interactions between large bipolarons result from their self-trapped electronic carriers adjusting to atoms' vibrations. As described in Chapter 4, the adjustment of a self-trapped carrier to vibrations of the associated atoms reduces the related stiffness constants thereby lowering the vibrations' energy. Moreover, the net lowering of the vibration energy produced by a collection of large bipolarons increases as their mutual separations decrease (Emin, 1994a; 1994b). In particular, a second-order perturbation calculation of the energy shift produced by the softening arising from a collection of large bipolarons depends on their mutual separations. This dependence arises as the self-trapped carriers respond coherently to vibrations whose wavelengths exceed the separation between large bipolarons. The change of the zero-point vibration energy induced by two large bipolarons of radius R_b separated by the distance s is roughly $-\hbar\omega(R_b/s)^4$. This energy lowering describes an attractive force between large bipolarons. Here, as in the BCS theory of metals' superconductivity, the ground-state energy is reduced by electrons' coherent response to zero-point atomic vibrations. In both instances the magnitude of this energy lowering increases as atoms' vibration frequencies are increased.

14.2 Condensation to a large-bipolaronic liquid

Driven by attractive inter-bipolaron interactions, large bipolarons may condense into a liquid phase in an analogous manner to that by which ^{4}He atoms condense into a liquid phase. The zero-point phonon-mediated attractions between large bipolarons described in Section 14.1 then replace the direct van der Waals (electrostatic) attractions between ^{4}He atoms. These attractive interactions will overwhelm large-bipolarons' mutual Coulomb repulsions if they are sufficiently suppressed by a material's exceptionally large static dielectric constant. Indeed, as described in Section 7.1, a very large ratio of static-to-optic dielectric constants is required for large-bipolaron formation. Thus, an especially large static dielectric constant is a reasonable presumption in instances in which large bipolarons form. Then, as illustrated in Fig. 14.1, the sum of the three interactions described in the previous section produces a minimum of the net interaction potential V_{int} near where the inter-bipolaron separation s is comparable to twice the large-bipolaron radius R_b.

Establishing a minimum in the interaction potential insures that large bipolarons will condense at sufficiently low temperatures. Calculation of the associated phase diagram proceeds in analogy with that for the condensation of a gas into a van der Waals liquid. The curve in Fig. 14.2 shows the temperature below which a gas of large bipolarons begins condensing into a bipolaron liquid versus the bipolaron concentration (Emin, 1994a; 1994b). This condensation temperature T_c increases with the depth of the attractive potential while the van der Waals repulsion volume is only qualitatively related to that of a large bipolaron (Landau and Lifshitz, 1958). A vertical dashed line is added to Fig. 14.2 to schematically denote the carrier density

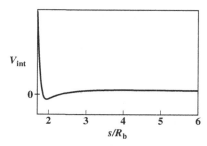

Fig. 14.1 When the static dielectric constant is sufficiently large the interaction potential V_{int} comprising the three components described in Section 14.1 develops a minimum at a separation between large bipolarons s which is near twice the bipolaron radius R_b.

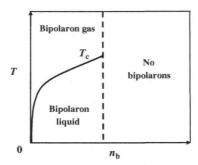

Fig. 14.2 Phonon-mediated attractions between large bipolarons enable them to condense from a gas of nearly independent quasi-particles into a liquid phase. Condensation begins when the temperature T is lowered below the condensation temperature T_c. The condensation temperature rises with the large-bipolaron density n_b. However, large bipolarons cannot form beyond a maximum limiting density depicted by the vertical dashed limit.

beyond which large bipolarons can no longer form. In particular, large bipolarons must be sufficiently sparse to provide enough polarizable medium about each of them to enable self-trapping.

The radius of a Fröhlich large-bipolaron R_b nearly equals that of a Fröhlich large-polaron R_p. In particular, as described in Section 7.1, the contraction of the self-trapped state induced as pairing doubles the depth of the self-trapping potential well offsets the expansion of the state induced by the pair's mutual Coulomb repulsion. The radius of a Fröhlich large polaron is $R_p \equiv (\hbar^2/me^2)[\varepsilon_\infty \varepsilon_0/(\varepsilon_0 - \varepsilon_\infty)]$, where m denotes the electronic carrier's effective mass while ε_0 and ε_∞, respectively, represent the material's static and optical dielectric constants. Evaluation of this formula indicates that R_p is usually not much larger than an interatomic separation. For instance, (1) $\hbar^2/me^2$ is just the Bohr radius (≈ 0.5 Å) if m equals the free-electron mass and (2) $\varepsilon_0 \geq 2\varepsilon_\infty$ for large bipolaron formation. By contrast, a small bipolaron is one that has collapsed to the smallest size acceptable in a medium composed of discrete atoms and bonds (e.g. just a single atom, ion or bond).

Figure 14.2 only depicts the phase boundary delineating where all large bipolarons exist as a gas and where large bipolarons first condense into a liquid. Moreover, as the temperature is raised large bipolarons in the gas may decompose into large polarons and self-trapped carriers may be liberated thereby spawning free carriers. Furthermore, cooling the liquid can drive some carriers to organize spatially in a manner commensurate with the underlying solid's lattice structure. These carriers may even collapse onto individual ions, atoms or bonds thereby forming regular arrays of small polarons or small bipolarons. That portion of the large-bipolaron liquid which survives as the temperature approaches absolute zero may participate in its superconducting condensate.

14.3 Superconductivity and excitations of a large-bipolaronic liquid

Large-bipolaronic superconductivity, like the superfluidity of ^{4}He, is attributed to collective behavior of mutually interacting particles whose ground state remains liquid. Liquidity persists in the ^{4}He ground state because the mutual attractive potential between ^{4}He atoms is so weak that the amplitude for their zero-point vibrations is of the order of the interatomic separation (Feynman, 1972). A similar situation prevails for a liquid of large bipolarons since their weak ($< \hbar\omega$) mutual attractive potential is comparable to their zero-point kinetic energy $\sim \hbar^2/m_b R_b^2 < \hbar\omega$. One sees that $\hbar^2/m_b R_b^2 \sim (\hbar\omega)^2/E_p < \hbar\omega$ upon recalling from Chapters 3, 4 and 10 that $m_b \sim E_p/\omega^2 R_b^2$ with the large-polaron's binding energy being large enough to insure self-trapping $E_p > \hbar\omega$.

Large-bipolarons' superconductivity is envisioned as analogous to the super-fluidity of ^{4}He. This approach presumes that singlet large bipolarons, like ^{4}He atoms, behave as bosons. The onset of superconductivity of a large-bipolaronic liquid, like the onset of superfluidity of liquid ^{4}He, is then taken to occur as the temperature is lowered below the liquid's Bose–Einstein condensation temperature. Below this condensation temperature the excitation-less ground state has a finite occupation even when reduced by the interactions between the condensate's parti-cles (Penrose and Onsager, 1956). The distinctive transport properties which col-lectively constitute superflow are attributed to this condensate. In particular, the condensate is characterized by resistance-less irrotational flow which yields the Meissner effect when the bosons are charged (Lifshitz and Pitaevskii, 1980).

The condensate's flow will be without resistance if it does not generate excitations. Excitations will not be generated if their creation is incompatible with conservation of energy and momentum. In particular, the Landau condition for resistance-less condensate flow is that its velocity v_c be less than the minimum value of $E(p)/p$, where $E(p)$ is the energy of a real (undamped) condensate excitation of momentum p (Landau and Lifshitz, 1958). The excitations for interacting bosons will be completely undamped if the curvature of the excitation spectrum is negative and momenta are kept below a threshold value (Pitaevskii, 1959; Abrikosov *et al.*, 1963; Lifshitz and Pitaevskii, 1980).

The excitations of the condensate depend on the interactions between its particles. The long-range Coulomb interactions between large bipolarons determine the long-wavelength limit of their excitation energy. As illustrated in Fig. 14.3, the long-wavelength limit of the excitation energy is simply $\hbar\omega_p$, where $\omega_p \equiv (4\pi n_b q^2/\varepsilon_0 m_b)^{1/2}$ is the plasma frequency for large bipolarons of density n_b, mass m_b and charge $q = 2e$ in a medium with static dielectric constant ε_0. Large bipolarons' mutual Coulomb repul-sions ensure that the Landau condition for resistance-less flow is satisfied at long wavelengths: $E(p)/p \rightarrow \hbar\omega_p/p \rightarrow \infty$ as $p \rightarrow 0$. By contrast, the Landau condition for

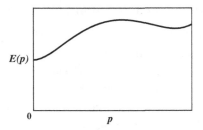

Fig. 14.3 The anticipated momentum dependence of the excitation energy $E(p)$ of a large-bipolaronic liquid is depicted. In the $p \to 0$ limit $E(p) \to \hbar\omega_p$ as long-range Coulomb interactions between large bipolarons predominate. The depiction of a ^{4}He-like "roton" minimum indicates that $E(p)$ may display complex behavior away from its $p \to 0$ limit.

resistance-less flow in liquid ^{4}He is satisfied at long wavelengths if the flow velocity v_c is less than the liquid's sound velocity s_1: $E(p)/p \to s_1 p/p = s_1 > v_c$.

The excitation energies of interacting large bipolarons are modest, only comparable to phonon energies. In particular, the hefty large-bipolaron effective mass, the large value of the static dielectric constant required for large-bipolaron formation ($\varepsilon_0 \gg \varepsilon_\infty > 1$), and the modest large-bipolaron density $n_b \ll R_b^{-3}$ result in $\omega_p < \omega$, the characteristic phonon frequency. In addition, the large-bipolaron's big effective mass restricts the overall spread of excitation energies to less than the phonon energy $\hbar\omega$.

14.4 Distinctive properties of a large-bipolaronic superconductor

Large-bipolaronic superconductivity involves singlet self-trapped carrier pairs which move very slowly within a medium whose exceptionally displaceable ions generate an especially large static dielectric constant $\varepsilon_0 \gg \varepsilon_\infty$. By contrast, BCS superconductivity involves fast-moving electronic charge carriers in a conventional metal. Some phenomena highlight these fundamental distinctions.

As described in Section 9.1, incident radiation is absorbed when a self-trapped carrier is excited within or from the potential well which binds it. Such absorptions generally produce phonon-broadened low-energy absorption bands. Absorption bands from exciting self-trapped carriers are produced by polarons and/or bipolarons that are (1) bound to dopants, (2) free of dopants and (3) participants in a super-conducting condensate. However, such distinctive absorption bands are not produced by electronic carriers in normal or superconducting states of conventional metals.

(Bi)polarons' photo-emission occurs when carriers that are photo-excited from their self-trapped states escape from the solid. The Franck–Condon principle ensures that atoms do not move during the excitation process. Thus this photo-emission is similar to that from a doped semiconductor whose carriers are bound in conventional traps. Moreover, below a bipolaronic superconductor's condensation temperature dispersion of the self-trapping energies will generally narrow as (bi) polarons enter their homogeneous condensate. Distinctively, with bipolaron formation the density of self-trapped states equals the density of self-trapped pairs. Thus, the electrons' chemical potential remains pinned between the energy of the self-trapped carriers and that of the edge of the solid's non-polaronic states. By contrast, the Fermi energy of a non-polaronic material shifts as the states' occupancies are altered.

Positrons injected into a solid can recombine with its electrons. Positron annihilation is evidenced by the emission of gamma rays. The positron annihilation rate depends on the electron density and its homogeneity. In particular, the positron lifetime is sensitive to the presence of localized electrons. Thus, a materials' normal-state positron lifetime should also fall upon introducing self-trapped carriers. However the positron lifetime ought to rise as the temperature is lowered below the condensation temperature where self-trapped carriers merge into the bipolaronic liquid's homogeneous superconducting condensate. By contrast, the positron lifetime of a conventional superconductor is not altered upon passing into the superconducting state.

The shifting of large (bi)polarons' self-trapped charges in response to atomic vibrations reduces their associated stiffness constants. If this carrier-induced softening is strong enough, multi-site self-trapped carriers of radius R_b induce local vibration modes having wavelengths $\leq R_b$ (Emin, 1991b). Thus, large (bi)polarons' presence can introduce short-wavelength vibration modes beyond those of the carrier-free solid. Such "extra" vibration modes have been called "ghost" modes.

All told, large (bi)polarons' self-trapped carriers produce multiple manifestations of localized charge carriers. Indeed, such charge localization can occur in the exceptional ionic materials in which large (bi)polarons can exist but not in conventional metals. Nonetheless, electronic localization by itself does not necessarily imply either self-trapping or large-bipolaron formation. Indeed, electronic localization results from trapping as well as from self-trapping.

Other distinctive large-(bi)polaronic properties directly address electronic transport. As described in Section 10.3, a large (bi)polaron possesses a large effective mass and is weakly scattered by phonons. By contrast, conventional metals' carriers have modest effective masses and are strongly scattered by phonons. In combination these two properties yield moderate mobilities for both large (bi)polarons and for conventional metals' carriers. However, measurements which individually address carriers' effective masses and scattering times distinguish between large (bi)polarons and metals' electronic carriers. In particular, scattering times can be deduced from the frequency dependence of the electrical conductivity. Scattering times for large (bi)polarons should approach atoms' characteristic vibration period. However, scattering times are generally much shorter for metals' charge carriers. Furthermore, as discussed in Section 10.4, a large (bi)polaron's scattering by acoustic phonons peaks for half-wavelengths comparable to the large (bi)polaron's diameter. The resulting scattering time and mobility are inversely proportional to the thermal energy κT when it exceeds the energies of these phonons.

Doping is frequently used to introduce charge carriers to the high-dielectric-constant ionic insulators that are candidates for supporting large-bipolaronic super-conductivity: e.g. $SrTiO_3$ and La_2CuO_4. However, the dopant density need not be directly related to the density of mobile carriers. In some cases the multiplicity of valences that constituents of these ionic materials can assume (e.g. Tl can be 1+ or 3+) introduces uncertainty in the relationship between the doping density and the carrier density. Furthermore, some carriers may join solidified phases in which carriers order as small or large polarons or bipolarons in a manner which is commensurate with a crystal's underlying lattice structure. Then only the residual non-solidified carriers can contribute to a large-bipolaronic liquid.

A large-bipolaronic liquid and its superconductivity will disappear when carriers order in a manner that is commensurate with the solid's underlying lattice (Emin, 1994a; 1994b). If paired carriers form a square super-lattice whose period is d times the underlying lattice constant, commensurate arrangements will occur at carrier concentrations equaling the simple fractions $2/d^2$. As illustrated in Fig. 14.4, this situation may occur in $A_xLa_{2-x}CuO_4$, where holes are introduced to the quasi-square CuO_2 planes of La_2CuO_4 by substituting divalent cations $A \equiv Ca$, Sr or Ba for the trivalent cation La. In particular, the proximity of the compositional limits of the superconducting phase of $A_xLa_{2-x}CuO_4$ to $x = 2/25$ and $x = 2/9$ with the loss of

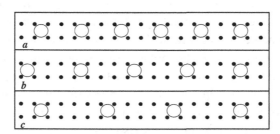

Fig. 14.4 Charge carriers in doped La_2CuO_4 are envisioned as residing in CuO_2 planes. Each plane is represented as a square array of Cu^{2+} cations (solid dots). To simplify this figure, oxygen O^{2-} anions, centered midway between each pair of nearest-neighbor Cu cations, are *not* shown. The core of each large bipolaron (open circles) is envisioned as encompassing the four contiguous oxygen anions bounded by the four Cu^{2+} cations of a square. At critical densities such large bipolarons can order commensurately with the underlying square lattice. In panels *a*, *b*, and *c* such ordering is depicted along one of the square lattice's principal axes when a large bipolaron is centered in every $3 \times 3 = 9$, $4 \times 4 = 16$ and $5 \times 5 = 25$ structural units, respectively.

Fig. 14.5 An eight-atom structural unit of a CuO_2 plane of an insulating cuprate contains four Cu^{2+} cations (solid dots) each with a net spin of ½ (arrow) and four contiguous O^{2-} anions (open circles). The shifting of cation and anion nuclei induced by adding two holes to the four-oxygen complex (dotted region), the core of the envisioned large-bipolaron, is hypothesized to transfer electrons from oxygen sites to copper sites leaving them with paired spins.

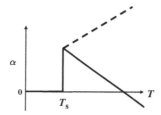

Fig. 14.6 The Seebeck coefficient α of a superconductor vanishes when the temperature T is lowered below the superconducting transition temperature T_s. When $T > T_s$ the Seebeck coefficient of a conventional p-type superconductor rises with increasing temperature (dashed line). If, however, large-bipolaron formation eliminates local magnetic moments, the Seebeck coefficient of a p-type cuprate superconductor may fall and even change sign as the temperature is raised (solid line).

superconductivity at the intermediate concentration $x = 2/16$ is consistent with the self-trapped carriers being paired.

The insulating parents of the cuprate superconductors, e.g. La_2CuO_4, are anti-ferromagnets with spin-½ Cu^{2+} cations. With progressive doping, cuprates' magnetism disappears. If creating self-trapped holes eliminates local moments, carriers' Seebeck coefficients garner the negative spin entropy contribution described in Section 12.4. The magnitude of this negative contribution to the Seebeck coefficient increases with rising temperature as the removed-moments' spin-alignment entropy increases. For definiteness, consider the hypothesis illustrated in Fig. 14.5 in which formation of a p-type singlet large bipolaron centered on four contiguous oxygen ions removes the local moments of four adjacent spin-½ Cu^{2+} ions. This singlet-bipolaron formation process is described by the chemical formula: 2 holes + $(O_4)^{8-} + 4\,Cu^{2+} \rightarrow (O_4)^{2-} + 4\,Cu^{1+}$. Since four sites with spin-½ are eliminated in this process, it will contribute $-4(\kappa/2e)\ln(2)$ to the Seebeck coefficient in the high-temperature paramagnetic limit. This negative contribution to the Seebeck coefficient can drive it to resemble the schematic depiction of Fig. 14.6.

The superconducting condensate of a conventional metal will permeate into an adjoining metal thereby producing a proximity effect. Such proximity effects are not expected of a large-bipolaronic condensate since conventional metals will not sustain (bi)polaron formation. However, a large-bipolarons' condensate could extend into or pass through an electronic insulator whose static and optical dielectric constants support large-bipolaron formation, $\varepsilon_0 \gg \varepsilon_\infty$. Thereby a large-bipolaronic superconductor could exhibit novel proximity effects.

15

Non-Ohmic hopping conduction and electronic switching

Driving a steady-state current of carriers hopping between inequivalent states generally alters sites' occupancies. This carrier redistribution grows as the temperature is lowered or the current is increased. Carriers' redistribution becomes significant when the hopping conduction becomes non-Ohmic.

This chapter begins by describing how the current-versus-voltage relationship is determined for steady-state hopping conduction among inequivalent states. As flow becomes arbitrarily weak (1) carrier redistribution becomes ignorable, (2) charge transport becomes Ohmic, and (3) hopping transport reduces to that of a resistor network. The remainder of the chapter focuses on the carrier redistribution that accompanies non-Ohmic hopping.

Current-driven carrier redistribution of charge carriers is explored through two examples. Section 15.2 addresses low-temperature hopping among inequivalent states of a disordered medium. Section 15.3 describes effects of carrier transfers between a small-polaron semiconductor and non-polaronic electrical contacts. Increasing the steady-state current augments the densities of slow-moving small polarons which accumulate near the semiconductor's interfaces with its contacts. A sufficiently strong current may drive the density of interfacial small polarons high enough to preclude their formation. This effect suggests a mechanism for reversibly switching much of the semiconductor into a relatively high-conductance state.

15.1 Formalism for steady-state hopping

The net electric current produced by carriers of charge q hopping between sites i and j is

$$I_{i,j} = q\left[f_i(1-f_j)R_{i,j} - f_j(1-f_i)R_{j,i}\right], \tag{15.1}$$

where f_i and f_j are the probabilities of sites i and j being occupied and $R_{i,j}$ and $R_{j,i}$ are the rates with which a carrier jumps from site i to site j and from site j to site i, respectively. A site's occupation probability can be expressed in terms of a function which resembles the Fermi distribution function:

$$f_i \equiv \frac{1}{\exp[(E_i - \mu_i)/\kappa T] + 1}$$

$$= \frac{1}{2}\exp\left(-\frac{E_i - \mu_i}{2\kappa T}\right)\operatorname{sech}\left(\frac{E_i - \mu_i}{2\kappa T}\right) = \frac{p_i F_i}{1 + p_i F_i}, \qquad (15.2)$$

where μ_i, the local quasi-electro-chemical potential of site i replaces μ_c, the electro-chemical potential of an equilibrated system. Here E_i denotes the energy of a carrier at site i in the presence of an applied electric field and κT represents the thermal energy. Following the final equality of Eq. (15.2), f_i is expressed in terms of the local quasi-fugacity $F_i \equiv \exp(\mu_i / \kappa T)$ and the corresponding probability factor $p_i \equiv \exp(-E_i / \kappa T)$. Analogously,

$$1 - f_j \equiv \frac{1}{1 + \exp\left[-\left(E_j - \mu_j\right)/\kappa T\right]}$$

$$= \frac{1}{2}\exp\left(\frac{E_j - \mu_j}{2\kappa T}\right)\operatorname{sech}\left(\frac{E_j - \mu_j}{2\kappa T}\right) = \frac{1}{1 + p_j F_j}. \qquad (15.3)$$

With thermally equilibrated phonons, the rate for a carrier's phonon-assisted hop from site i to site j assumes the general form (Emin, 1974)

$$R_{i,j} = \exp\left[-(E_j - E_i)/2\kappa T\right]r_{i,j}(|E_j - E_i|) = \sqrt{\frac{p_j}{p_i}}r_{i,j}(|E_j - E_i|), \qquad (15.4)$$

where $r_{i,j}(|E_j - E_i|) = r_{j,i}(|E_i - E_j|)$. This rate and that for the reverse process manifestly satisfy the requirement of detailed balance:

$$R_{j,i}/R_{i,j} = \exp\left[(E_j - E_i)/\kappa T\right] = p_i/p_j. \qquad (15.5)$$

With Eqs. (15.2)–(15.4) incorporated into Eq. (15.1) it becomes evident that current flow between a pair of sites depends upon the differences of their quasi-electro-chemical potentials and quasi-fugacities, $\mu_i - \mu_j$ and $F_i - F_j$, respectively:

$$I_{i,j} = \frac{q}{2}r_{i,j}\left(|E_j - E_i|\right)\mathrm{sech}\left(\frac{E_i - \mu_i}{2\kappa T}\right)\mathrm{sech}\left(\frac{E_j - \mu_j}{2\kappa T}\right)\sinh\left(\frac{\mu_i - \mu_j}{2\kappa T}\right)$$

$$= qr_{i,j}\left(|E_j - E_i|\right)\frac{\sqrt{p_i p_j}}{\left(1 + p_i F_i\right)\left(1 + p_j F_j\right)}\left(F_i - F_j\right). \tag{15.6}$$

With steady-state flow, the carrier density at each site becomes time independent as the net current flowing to it vanishes:

$$\sum_j I_{i,j} = 0, \tag{15.7}$$

for each site labeled by i. Equation (15.7) can be solved for the sets of μ_i or of F_i, where the values of these quantities at the boundaries of a sample are determined by the emf imposed across it. The quasi-electro-chemical potential and quasi-fugacity at each site depend on the site energies and their shifts with the applied electric field. With these values Eq. (15.6) can be used to obtain the current between any pair of sites as a function of the applied field. Of course, no current flows in equilibrium since the quasi-electro-chemical potential and the quasi-fugacity at every site then revert to their equilibrium values, μ_c and $F_c \equiv \exp(\mu_c/\kappa T)$, respectively.

When the applied electric field and the associated inter-site differences of the quasi-electro-chemical potentials are sufficiently small, $\mu_i - \mu_j \ll \kappa T$, the currents of Eq. (15.6) become simply proportional to these differences (Miller and Abrahams, 1960). The expression after the first equality in Eq. (15.6) then reduces to Ohm's law: $I_{i,j} = G_{i,j}(\mu_i - \mu_j)$, with the conductance

$$G_{i,j} \equiv \frac{q}{4\kappa T}r_{i,j}\left(|E_j^0 - E_i^0|\right)\mathrm{sech}\left(\frac{E_i^0 - \mu_c}{2\kappa T}\right)\mathrm{sech}\left(\frac{E_j^0 - \mu_c}{2\kappa T}\right), \tag{15.8}$$

where the superscript 0 indicates local energies in the absence of the applied electric field. If the differences of the quasi-electro-chemical potentials for all significant hops of a system satisfy this condition, its hopping can be treated as a resistance network. However, this condition cannot be fulfilled if the temperature is too low or the current is too strong. The remainder of this chapter will discuss examples with such non-Ohmic behavior.

15.2 Low-temperature hopping in a disordered medium

Steady-state current flow drives a redistribution of sites occupied by hopping charge carriers. This effect becomes especially pronounced at low temperatures. To illustrate this effect and its implications for charge transport, consider just the current produced by a low density of charge carriers hopping between nearest neighbors along a linear chain of N sites.

With the presumption of a low carrier density, $p_i F_i \ll 1$, the expression for the current flowing between a pair of sites, Eq. (15.6), reduces to (Emin, 2006b)

$$I_{i,j} = q r_{i,j} \left(|E_j - E_i| \right) \sqrt{p_i p_j} \left(F_i - F_j \right). \tag{15.9}$$

Using this equation the condition for steady-state nearest-neighbor hopping at each interior chain site, $2 \le i \le N-1$, Eq. (15.7) becomes

$$\frac{F_{i-1} - F_i}{F_i - F_{i+1}} = \frac{r_{i,i+1}(|E_{i+1} - E_i|)}{r_{i-1,i}(|E_i - E_{i-1}|)} \sqrt{\frac{p_{i+1}}{p_{i-1}}} \equiv a_i. \tag{15.10}$$

The positive sign of the right-hand side of this equation for all values of i implies that the quasi-fugacity F_i changes monotonically along the chain. The quasi-fugacities at the ends of the chain, F_1 and F_N, are established by the boundary conditions for steady-state flow: the zero of potential energy and the imposed emf.

The quasi-fugacity at an intervening site s along the chain is determined by iterative solution of Eq. (15.10) to be (Emin, 2006b):

$$F_s = \frac{F_1 G_s + F_N (G_1 - G_s)}{G_1} \tag{15.11}$$

with $G_N \equiv 0$, where

$$G_s \equiv 1 + \sum_{i=s+1}^{N-1} \prod_{j=i}^{N-1} a_j \tag{15.12}$$

for $1 \le s < N$. The steady-state current for nearest-neighbor hopping along the chain is found by inserting Eqs. (15.11) and (15.12) into Eq. (15.9):

$$I = I_{N-1,N} = \frac{q r_{N-1,N}(|E_N - E_{N-1}|) \sqrt{p_N p_{N-1}}}{1 + \sum_{i=2}^{N-1} \prod_{j=i}^{N-1} a_j} \left(F_1 - F_N \right). \tag{15.13}$$

The quasi-fugacity difference $F_1 - F_N$ drives the current. There are $N-1$ contributions to the denominator of Eq. (15.13). In the Ohmic limit each contribution is proportional to the corresponding resistance of an $(N-1)$-link chain of resistors.

Beyond the Ohmic limit, the electric-field dependences of factors on the right-hand side of Eq. (15.13) other than $F_1 - F_N$ are no longer ignorable.

Attention is now focused on low-temperature hopping along the one-dimensional chain. In the low-temperature limit the net energy difference for a phonon-assisted hop upward in energy becomes the jump rate's activation energy. Meanwhile the rate for a hop downward in energy becomes temperature independent since it occurs via the spontaneous emission of phonons (Miller and Abrahams, 1960; Emin, 1974; 1975a; Bergeron and Emin, 1977). These two features emerge from the low-temperature limit of the general expression for a phonon-assisted jump rate Eq. (15.4) c.f. Eq. (11.20):

$$\lim_{T \to 0} r_{i,j}\left(\left|E_j - E_i\right|\right) \propto \exp\left(-\left|E_j - E_i\right|/2\kappa T\right). \tag{15.14}$$

The remaining factors in $r_{i,j}(|E_j - E_i|)$ depend on the specifics of the electron–phonon interaction. Indeed, the low-frequency ac conductivity of adiabatic small-polaron hopping at low temperatures, dominated by hops downward in energy, is determined by these dependences (Emin, 1992). Nonetheless, these dependences have been disregarded when addressing the low-temperature limit of the dc conductivity which is dominated by hops upward in energy (Ambegaokar *et al.*, 1971). With this approximation, $r_{i,j}(|E_j - E_i|) = r_0 \exp(-|E_j - E_i|/2\kappa T)$ with r_0 being just a constant. Upon adopting this simplification the site coefficients defined in Eq. (15.10) become

$$a_i = \frac{e^{-\left|E_{i+1}-E_i\right|/2\kappa T}}{e^{-\left|E_i-E_{i-1}\right|/2\kappa T}} \frac{e^{-E_{i+1}/2\kappa T}}{e^{-E_{i-1}/2\kappa T}}. \tag{15.15}$$

Evaluating Eq. (15.15) for the four possible variations of the energy with site index gives: (1) $a_i = \exp[-(E_{i+1} - E_i)/\kappa T]$ for $E_{i-1} < E_i < E_{i+1}$, (2) $a_i = 1$ for $E_{i-1} < E_i > E_{i+1}$, (3) $a_i = \exp[-(E_i - E_{i-1})/\kappa T]$ for $E_{i-1} > E_i > E_{i+1}$, and (4) $a_i = \exp[-(E_{i+1} - E_{i-1})/\kappa T]$ for $E_{i-1} > E_i < E_{i+1}$.

Lowering the temperature or increasing the steady-state current produced by carriers' hopping among energetically inequivalent sites alters their occupancies. Concomitantly the relationship between the current and its electromotive force becomes increasingly nonlinear. The simplest example of hopping along a linear chain is used to illustrate these features.

Figure 15.1 depicts a succession of four hops which transport a carrier over an energy barrier. The energies of the first and fifth sites of the five-site chain are chosen to equal one another in the absence of the applied field. Since the energy of a carrier includes the potential energy of the imposed electric field, the difference between the energy of a carrier at site 1 and at site 5 is just the product of the applied potential V and the carrier charge q: $E_1 - E_5 = qV$.

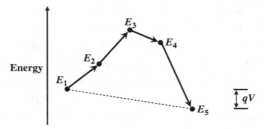

Fig. 15.1 This illustration depicts the energies for the four nearest-neighbor hops via which the emf qV drives a carrier over an energy barrier on a five-atom linear chain.

Imposing the electromotive force establishes the boundary conditions on the quasi-fugacities at sites 1 and 5: $F_1 = 1$ and $F_5 = \exp(-qV/\kappa T)$. The quasi-fugacities at sites 2, 3, and 4 are found from Eqs. (15.11) and (15.12). Then the model of Eq. (15.15) is utilized to express the a_i in terms of the E_i. The results are:

$$F_2 = \frac{F_5(a_2 a_3 a_4) + F_1(a_3 a_4 + a_4 + 1)}{(a_2 a_3 a_4 + a_3 a_4 + a_4 + 1)}$$

$$= \frac{e^{-qV/\kappa T} e^{(E_2 - E_4)/\kappa T} + 2e^{(E_3 - E_4)/\kappa T} + 1}{e^{(E_2 - E_4)/\kappa T} + 2e^{(E_3 - E_4)/\kappa T} + 1}, \qquad (15.16)$$

$$F_3 = \frac{F_5(a_2 a_3 a_4 + a_3 a_4) + F_1(a_4 + 1)}{(a_2 a_3 a_4 + a_3 a_4 + a_4 + 1)}$$

$$= \frac{e^{-qV/\kappa T}\left[e^{(E_2 - E_4)/\kappa T} + e^{(E_3 - E_4)/\kappa T}\right] + e^{(E_3 - E_4)/\kappa T} + 1}{e^{(E_2 - E_4)/\kappa T} + 2e^{(E_3 - E_4)/\kappa T} + 1}, \qquad (15.17)$$

and

$$F_4 = \frac{F_5(a_2 a_3 a_4 + a_3 a_4 + a_4) + F_1}{(a_2 a_3 a_4 + a_3 a_4 + a_4 + 1)}$$

$$= \frac{e^{-qV/\kappa T}\left[e^{(E_2 - E_4)/\kappa T} + 2e^{(E_3 - E_4)/\kappa T}\right] + 1}{e^{(E_2 - E_4)/\kappa T} + 2e^{(E_3 - E_4)/\kappa T} + 1}. \qquad (15.18)$$

Figure 15.2 depicts the quasi-electro-chemical potential for three distinct limiting regimes. (1) When $\kappa T \gg E_3 - E_4, E_4 - E_2, qV$, the quasi-electro-chemical potential $\mu_i \equiv \kappa T \ln(F_i)$ varies almost uniformly along the chain. Then, as depicted by the dotted line in Fig. 15.2, $\mu_1 = 0$, $\mu_2 = -qV/4$, $\mu_3 = -qV/2$, $\mu_4 = -3qV/4$, $\mu_5 = -qV$. (2) As the temperature is lowered the thermal energy falls below the electronic energy

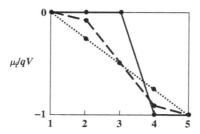

Fig. 15.2 The quasi-electro-chemical potentials μ_i in units of the driving emf qV are plotted at the five sites (large dots) involved in flow across the energy barrier depicted in Fig. 15.1. Three situations are described. (1) The values at the five sites are linked by the dotted line when the thermal energy κT greatly exceeds both the barrier height and the applied emf. (2) The values at the five sites are linked by the dashed line when κT is much less than the barrier height but much greater than qV. (3) The five sites are linked by the solid line in the low-temperature limit in which κT is much less than both the barrier height and qV.

differences, $qV \ll \kappa T \ll E_4 - E_2, E_3 - E_4$. Then, as shown by the dashed curve of Fig. 15.2, the variation of the quasi-electro-chemical potential occurs primarily in the vicinity of the barrier which limits carriers' flow. In particular, the dashed line in Fig. 15.2 shows the quasi-electro-chemical potentials for the energy barrier depicted in Fig. 15.1. The corresponding values are: $\mu_1 = 0$, $\mu_2 \approx 0$, $\mu_3 \approx -qV/2$, $\mu_4 \approx -qV$, $\mu_5 = -qV$. (3) Finally, as the temperature is lowered toward absolute zero $\kappa T \ll qV \ll E_4 - E_2, E_3 - E_4$. Then, as shown by the solid line of Fig. 15.2, the variation of the quasi-electro-chemical potential occurs primarily over the downstream portion of the barrier which limits carriers' flow. In particular, the solid line in Fig. 15.2 shows the low-temperature limits of the quasi-electro-chemical potentials for the energy barrier depicted in Fig. 15.1. The corresponding values are: $\mu_1 = 0$, $\mu_2 \approx 0$, $\mu_3 \approx -\kappa T \ln(2)$, $\mu_4 \approx -qV$, $\mu_5 = -qV$.

Steady-state current flow alters sites' occupation probabilities. The occupation probability of a site f_i relative to its equilibrium value f_i^0 is

$$f_i/f_i^0 = \exp\{[qV(i-1)/4 + \mu_i]/\kappa T\}. \tag{15.19}$$

Figure 15.3 displays $\ln(f_i/f_i^0)/(qV/\kappa T)$ plotted against the site index i for the three limits depicted in Fig. 15.2. As the temperature is lowered carriers participating in the flow illustrated in Fig. 15.1 increasingly accumulate on the upstream side of the energy barrier and are depleted on its downstream side.

The steady-state current through the energetic landscape depicted in Fig. 15.1 is obtained from Eq. (15.13) after writing the quasi-fugacities F_1 and F_5 in terms of the electromotive force and using the model of Eq. (15.15) to express the a_i coefficients in terms of the field-dependent carrier energies, the E_i.

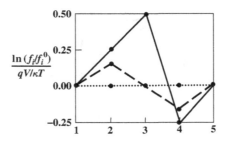

Fig. 15.3 The logarithm of the ratio of a site's occupation probability under steady-state flow f_i to that in equilibrium f_i^0 is plotted in units of the ratio of the emf to the thermal energy $qV/\kappa T$ for the five sites comprising the energy barrier of Fig. 15.1. As the temperature is lowered carriers increasingly accumulate in front of the barrier while being depleted behind it.

$$I = \frac{qr_0}{e^{E_2/\kappa T} + 2e^{E_3/\kappa T} + e^{E_4/\kappa T}}\left(1 - e^{-qV/\kappa T}\right)$$

$$\rightarrow qr_0 e^{-E_3^0/\kappa T}\sinh(qV/2\kappa T). \tag{15.20}$$

Here E_3^0 is the energy of the carrier at site 3 in the absence of the applied electric field and r_0 is the jump-rate constant introduced above Eq. (15.15). The expression after the arrow in Eq. (15.20) indicates that the current's low-temperature limit only depends on the energy-landscape's highest field-free energy E_3^0.

This current becomes non-Ohmic in the low-temperature limit $T \rightarrow 0$ while keeping V finite. Nonetheless, the low-temperature limit of a disordered medium's hopping conductivity was addressed under the presumption that it remains Ohmic, $I \propto V$ (Ambegaokar *et al.*, 1971). In particular, hopping transport was assumed to be describable by a linear-response resistor network with conductances given by Eq. (15.8). Such a limit corresponds to taking $T \rightarrow 0$ while also requiring that $\kappa T \gg qV$.

The one-dimensional example of this section illustrates how steady-state flow of hopping charge carriers in a disordered medium alters the probabilities of sites being occupied. This effect increases as (1) the temperature is lowered, (2) the current is raised and (3) the disorder is enhanced. Increasing the dimensionality quantitatively reduces the effects of disorder by offering more ways for carriers to avoid negotiating especially difficult jumps.

15.3 Interfacial small-polaron accretion: a threshold switching mechanism

Driving a steady-state current of small polarons can enhance their densities near its semiconductor's interfaces with high-mobility contacts. Two effects specific to small-polarons foster their accumulation near these interfaces. First, the mobilities of small polarons are generally much lower than those of carriers moving through conventional leads. Second, conversion between the semiconductor's small polarons and the leads' conventional carriers is slowed by the need to move the atoms associated with small-polaron formation. As the current is increased small-polarons' interfacial densities rise toward a maximum beyond which they can no longer form. Reaching this maximum is posited to trigger an avalanche instability in which all but a residual region of the semiconductor reversibly switches into a state of higher conductance.

A narrow-band model is utilized to illustrate the redistribution of carriers that accompanies their current-driven passages between high-mobility leads and a low-mobility small-polaron semiconductor (Emin, 2006b). As illustrated in Fig. 15.4 a linear chain of N sites is divided into a central portion of n small-polaron sites and two m-site leads, $N = n + 2m$. Disorder-related energy variations among each of the two classes are incidental and are therefore ignored. The energy at a small-polaron site labeled by i is then $-E_b - qV(i-1)/(N-1)$, where E_b is the small-polaron binding energy and $V/(N-1)$ is the drop of the applied potential between adjacent sites. Analogously, the energy at a lead site is $-E_f - qV(i-1)/(N-1)$, where E_f is a free-carrier's energy in the absence of the applied electric field.

Carriers are presumed to move only between adjacent sites on the chain. The rates for these inter-site transitions are written in the general form displayed in Eq. (15.4). However transfer processes involving the semiconductor and its leads are characterized by distinct rate factors $r_{i,i\pm1}(|E_{i\pm1}-E_i|)$. For phonon-assisted hops of a carrier between small-polaron sites $r_{i,i\pm1}(|E_{i\pm1}-E_i|) \equiv R_s$. For a carrier transferring between neighboring lead sites $r_{i,i\pm1}(|E_{i\pm1}-E_i|) \equiv R_l$. For a carrier moving from a small-polaron site to a neighboring lead site $r_{i,i\pm1}(|E_{i\pm1}-E_i|) \equiv R_{s,l}$. Similarly, $r_{i,i\pm1}(|E_{i\pm1}-E_i|) \equiv R_{l,s}$ for a carrier moving from a lead site to an adjacent small-polaron site.

The steady-state current I and the concomitant shifts of sites' occupancies f_i/f_i^0 are functions of these rate factors and the driving potential V. The procedure to establish these dependences becomes especially simple if the site occupancies are presumed small. Then I and f_i/f_i^0 are functions of every site's steady-state coefficient a_i and V with Eq. (15.10) expressing each a_i in terms of the rate factors and V.

The values of a_i are obtained from Eq. (15.10) for the model of Fig. 15.4. For the left-side lead, $2 \leq i \leq m-1$, $a_i = \exp[qV/(N-1)\kappa T] \equiv A$. For the two sites bridging the

$$\begin{array}{ccccc} 1 & & m & m+n & N=n+2m \end{array}$$

Fig. 15.4 A line of N sites is divided into two sets of m sites (solid dots) which represent leads to a semiconductor of n sites (open dots) upon which small polarons can form.

semiconductor's left interface with a lead, $a_m = A(R_{1,s}/R_1)\exp(-\Delta/2\kappa T)$ and $a_{m+1} = A(R_s/R_{1,s})\exp(-\Delta/2\kappa T)$, where $\Delta \equiv E_f - E_b$. For the semiconductor's interior sites, $m+2 \le i \le m+n-1$, $a_i = A$. For the two sites bridging the semiconductor's right interface with a lead, $a_{m+n} = A(R_{s,1}/R_s)\exp(\Delta/2\kappa T)$ and $a_{m+n+1} = A(R_1/R_{s,1})\exp(\Delta/2\kappa T)$ with $R_{s,1} = R_{1,s}$. For the right-side lead, $m+n+2 \le i \le N-1$, $a_i = A$. Thus the a_i-coefficients only depart from A at interfacial sites.

These modifications of interfacial sites' a_i-coefficients cause the probability of a site s being occupied, f_s, to depart from its equilibrium value f_s^0 under steady-state flow. The deviation of f_s/f_s^0 from unity results from a competition between the electric-field-dependences of the site's energy and its quasi-fugacity F_s (Emin, 2006b):

$$\frac{f_s}{f_s^0} = \exp\left[\frac{qV(s-1)/(N-1)}{\kappa T}\right]F_s = A^{s-1}F_s. \tag{15.21}$$

With the boundary conditions corresponding to imposing the emf qV, $F_1 = 1$ and $F_N = \exp(-qV/\kappa T) = A^{1-N}$, the quasi-fugacity at site s given by Eq. (15.11) becomes:

$$F_s = \frac{G_s + A^{1-N}(G_1 - G_s)}{G_1}. \tag{15.22}$$

Combining Eq. (15.22) with Eq. (15.21) yields

$$\frac{f_s}{f_s^0} = \frac{A^{s-1}G_s + A^{s-N}(G_1 - G_s)}{G_1}, \tag{15.23}$$

where G_s is obtained by inserting the coefficients determined in the preceding paragraph into Eq. (15.12).

This procedure, analogous to the calculation of Sec. 15.2, has been performed for this section's model (Emin, 2006b). Once the leads are long enough so that their lengths do not affect the accretion and depletion of small polarons within the semiconductor:

$$\frac{f_u}{f_u^0} = A^{u-n-2} + \left(\frac{R_l}{R_s}e^{\Delta/\kappa T}\right)\left[1-A^{u-n-1}-\delta_{u,1}\left(1-A^{-1}\right)\right]$$
$$+ \left(\frac{R_l}{R_{s,l}}e^{\Delta/2\kappa T}\right)\left(1-A^{-1}\right)\left(A^{u-n-1}+\delta_{u,1}\right), \tag{15.24}$$

where the n semiconductor sites are now labeled by u, $n \ge u \ge 1$. Small polarons' site occupations revert to their equilibrium values, $f_u \to f_u^0$, in the limits that (1) the applied electric field vanishes, $A \rightarrow 1$, or (2) interfacial disruptions of flow vanish, $R_s = R_1 = R_{s,1}$ with $\Delta = 0$.

Small-polarons' accumulation within the semiconductor are driven by their relatively slow transfers (1) through it, $R_s \ll R_1$, and (2) with its leads, $R_{s,1} \ll R_1$. As seen from Eq. (15.24) these impediments to flow act like the erecting of barriers produced when the semiconductor's energy levels rise above those of the leads to give $\Delta > 0$.

The contribution to the increase of the small-polaron density which is proportional to R_1/R_s in Eq. (15.24) rises from its value at the incident interface, site 1, to a peak at site 2 and then falls as the exiting interface is approached. By contrast, the contribution to the increase of the small-polaron density which is proportional to $R_1/R_{s,1}$ in Eq. (15.24) falls from its value at the incident interface to a minimum at site 2 and then rises as the exiting interface is approached. This interfacial-transfer contribution gains in prominence at high fields due to its proportionality to the factor $(1-A^{-1}) \equiv 1- \exp[-qV/(N-1)\kappa T]$. Indeed, this contribution will ultimately dominate at high fields if $R_{s,1}\exp(\Delta/2\kappa T) \ll R_s$. Then, as illustrated in Fig. 15.5, increasing the emf can drive large accumulations of small polarons near the semiconductor's interfaces. The dashed curves of Fig. 15.5 show that this accretion of small polarons persists as the semiconductor's thickness is decreased. All told, high electric fields can drive large enhancements of a semiconductors'

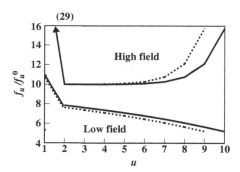

Fig. 15.5 The probability of site u on an n-site semiconductor chain being occupied during steady-state flow f_u relative to its equilibrium value f_u^0 is plotted against the site index u for carriers that enter the chain at site 1 and exit it from site n, $1 \le u \le n$. The two solid curves display f_u/f_u^0 for currents driven through a chain of $n = 10$ sites by high and low electric fields. These high- and low-field cases correspond to having potential differences between adjacent sites of κT and $\kappa T/10$, $\ln A = 1$ and 0.1 in Eq. (15.24), respectively. The two dashed curves are equivalent plots with the chain length reduced to $n = 9$. Carriers accumulate near the exiting interface when interfacial transfers between the semiconductor and the leads to it are impeded; here $(R_1/R_{s,1})\exp(\Delta/2\kappa T) = 40$ and $(R_1/R_s)\exp(\Delta/\kappa T) = 10$, where these physical parameters are defined in the text.

small-polaron density near their interfaces with high-mobility leads when charge transfers between them are severely restricted.

The rate associated with converting a carrier in a high-mobility lead into a small polaron in the semiconductor $R_{l,s}$ is reduced by the self-trapping of its carrier. As described in Section 4.3 and depicted in Fig. 4.3, a carrier introduced to a three-dimensional covalent material generally encounters a barrier to its self-trapping. Similarly, the rate associated with converting a small polaron in the semiconductor into a quasi-free carrier in a lead $R_{s,l}$ requires liberating the self-trapped carrier from the potential well which binds it. Such freeing of self-trapped carriers is slowed by the need to move the associated atoms. Thus conversions between quasi-free carriers in leads and small polarons in semiconductors are generally very slow.

Small-polaron semiconductors generally possess much larger densities of their low-mobility carriers than those of conventional semiconductors with their low densities of high-mobility carriers. Because of the slow conversion rates between a small-polaron semiconductors' low-mobility self-trapped carriers and the relatively high mobilities of leads' conventional carriers, strongly driven currents elevate small-polaron densities even higher.

However there is an upper limit to the permissible small-polaron density. In particular, as small-polarons' density is increased their carrier-induced deformations tend to interfere destructively with one another to limit the density for which stable small-polarons' formation is possible. The details of this interference depend on the specifics of the electron–phonon interactions, the phonons with which carriers interact and the material's structure (Emin, 1980b).

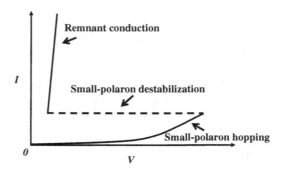

Fig. 15.6 An idealized *I*-versus-*V* curve for a semiconductor which manifests threshold switching shows a jump from low conductance (lower solid curve) to high conductance (upper solid curve) when flow exceeds a threshold. Reversion to low conductance occurs when high-conductance flow is sufficiently weakened. The model advanced here attributes the low-conductivity transport to the low-mobility hopping of a high density of small polarons. However, small polarons become destabilized when strong flow drives their density too high. The semiconductor then manifests the high conductance of the remnant.

Once the field is strong enough to drive an interfacial region's small-polaron density beyond its upper limit the region's carriers will no longer be self-trapped. Rather, the region's carriers will possess the greatly augmented mobility associated with quasi-free carriers. Thus the semiconductor's interfacial region will be converted into a free-carrier region. The net effect of this transformation just reduces the length of the semiconductor in which small polarons might form. However, Fig. 15.5 shows that simple shortening of the semiconductor shifts, but does not relieve, the field-driven elevation of the small-polaron density. This observation implies that a non-Ohmic strongly driven current will ultimately trigger an avalanche in which the semiconductor's carriers no longer form small polarons. As depicted in Fig. 15.6 this small-polaron destabilization provides a mechanism for threshold switching (Emin, 2006b).

Such switching phenomena are reported in many low-mobility solids in which small-polaron hopping is suspected, such as amorphous boron, transition-metal oxide glasses and chalcogenide glasses (Adler, 1971). For example, chalcogenide glasses manifest the high-temperature thermally activated mobilities and Hall-effect sign anomalies characteristic of small-polaron hopping (Emin *et al.*, 1972; Seager *et al.*, 1973; Seager and Quinn, 1975; Baily and Emin, 2006a; Baily *et al.*, 2006b).

The flow-induced destabilization of small polarons is put forth as a general mechanism for threshold switching in small-polaron semiconductors. However, this mechanism by itself is insufficient to explain the many intricacies observed for threshold switching of real non-crystalline films. For example, the relatively high conductance of the switched state of a chalcogenide glass is generally independent of the nominal length of its conducting path (Adler *et al.*, 1978; Fritzsche, 2006). In other words, the remnant conductance is not a bulk property of the switched glass.

16

Electronically stimulated desorption of atoms from surfaces

Foreign atoms can be adsorbed on the surface of a crystalline semiconductor. For example, hydrogen atoms can be adsorbed on a silicon crystal's surface. Irradiations with high-energy electrons, photons and even ions are observed to induce the expulsion of such adsorbed atoms. Observations support the notion that desorption from crystals is fundamentally different from that from small molecules where electronic excitations are necessarily localized (Jennison and Emin, 1983a).

Irradiation of an ideal periodic surface can produce delocalized surface electronic excitations. The time that such an excitation can dwell in the vicinity of an adsorbed atom is $\sim\hbar/W_e$, where W_e is the electronic excitation's bandwidth. This time is usually much less than the time required to significantly displace an atom $\sim 1/\omega$, where ω is the relevant atomic-vibration frequency. Thus, delocalized electronic excitations are generally unable to displace adsorbed atoms and thereby cleave their bonds to surface atoms.

Electronically stimulated desorption of an adsorbed atom is therefore associated with the severe localization of an electronic surface excitation which ruptures the bond to the adsorbed atom. This surface localization is a two-dimensional extension of that associated with small-polaron formation within solids (Jennison and Emin, 1983b). While small-polaron formation only envisions straining atoms' bonds, desorption also involves bond scission.

The adiabatic approach to self-trapping phenomena, described in Sections 4.1–4.3, is justified when electronic excitations' inter-site transfers are much faster than atoms' vibrations. Thus this approach was also adopted in developing a model of electronically stimulated desorption (Jennison and Emin, 1983b). As illustrated in Fig. 16.1, this desorption process envisions the progressive localization of the electronic excitation with concomitant displacements of adsorbed atoms until an adsorbed atom is expelled as its bond to the semiconductor surface is broken. Two types of desorption

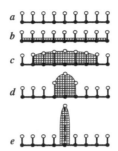

Fig. 16.1 The adiabatic treatment of electronically stimulated desorption is schematically illustrated. Panel *a* depicts absorbed atoms (open circles) bonded to a surface of a crystal's atoms (solid dots). Panel *b* depicts the introduction of a delocalized surface excitation. As atoms' equilibrium positions are progressively displaced ($c \rightarrow e$) the electronic excitation's localization increases until the bond to the desorbed atom is finally broken.

can occur. "Immediate desorption" proceeds unimpeded following irradiation. "Delayed desorption" is impeded by the need to traverse an energy barrier.

16.1 Model Hamiltonian

Electronically stimulated atomic desorption from a crystal surface occurs as severe localization of an electronic excitation culminates in breaking an adsorbed atom's bond to the surface. The localization process is analogous to that described in Chapter 4 for the collapse of a charge carrier during small-polaron formation. However, to describe desorption the electron–phonon interaction must include the bond-breaking which occurs as an adsorbed atom is expelled from the semiconductor's surface. Electronically induced bond-breaking is a limit of electronically induced softening. As discussed in Chapter 4, carrier-induced softening is a general feature of polaron formation. Softening occurs because a self-trapped carrier's wavefunction generally changes its size, shape, and centroid as associated atoms vibrate about their displaced equilibrium positions. This effect has frequently been ignored. Indeed, Holstein's molecular-crystal model does not even permit its small polaron to produce carrier-induced softening; its carrier's single-site wavefunction is implicitly represented as just a fixed point (Holstein, 1959a).

In the adiabatic limit atoms' motions become infinitesimally slow as their masses are imagined to become arbitrarily large. The ground-state energy is then just the minimum of the adiabatic potential. The adiabatic potential is the sum of (1) the strain energy associated with displacing adsorbed atoms from their equilibrium positions and (2) the energy of the electronic excitation as a function of these atomic displacements.

Treating the adsorbed atoms bonded to atoms on a covalent semiconductor's surface as a deformable continuum based on elementary structural units of volume V_c, the adiabatic potential becomes

$$V_{ad} = \frac{k}{2V_c}\int du \Delta^2(u) + \frac{\hbar^2}{2m}\int dr|\nabla_r\phi(r)|^2$$

$$+ \int dr|\phi(r)|^2 \int du \left[F\Delta(u) - \frac{k'}{2}\Delta^2(u)\right]\delta(r-u). \qquad (16.1)$$

The first term in Eq. (16.1) is the Hooke's law strain energy for adsorbed atoms bound to the semiconductor surface with stiffness constant k. Here $\Delta(u)$ denotes the displacement from equilibrium of an adsorbed atom whose nominal location is labeled by u. The second contribution in Eq. (16.1) denotes the kinetic energy of an added electronic excitation having wavefunction $\phi(r)$ and effective mass m. The final contribution in Eq. (16.1) is the expectation value of the electronic excitation's potential as a function of adsorbed atoms' displacements. The first contribution to this term, proportional to $\Delta(u)$, shifts adsorbed atoms' equilibrium positions by amounts proportional to F. The second contribution to this term, proportional to $\Delta^2(u)$, reduces the stiffness with which each adsorbed atom is bound to the semiconductor surface by an amount proportional to k'. The Dirac delta-function $\delta(r-u)$ expresses the short range of a covalent semiconductor's electron–phonon interaction. With $k' = k$ a severely localized electronic excitation removes the bond between the semiconductor's surface and an adsorbed atom thereby enabling its desorption.

The minimum of the adiabatic potential occurs when its first derivative with respect to $\Delta(u)$ vanishes, cf. Section 4.2:

$$\Delta(u) = -\frac{FV_c|\phi(u)|^2}{\left[k - k'V_c|\phi(u)|^2\right]}. \qquad (16.2)$$

After some algebraic manipulations the corresponding minimum of the adiabatic potential is found to be

$$V_{ad}^{min} = \frac{\hbar^2}{2m}\int dr|\nabla_r\phi(r)|^2 - \left(\frac{F^2}{2k}\right)V_c\int dr\frac{|\phi(r)|^4}{\left[1 - \left(\frac{k'}{k}\right)V_c|\phi(r)|^2\right]}. \qquad (16.3)$$

This result can be compared with Eq. (4.29) with only its short-range electron–phonon interaction. Electronically induced softening of bonds, manifested by $k' > 0$

in the second term of Eq. (16.3), fosters electronic localization by lowering the adiabatic potential's minimum.

16.2 Scaling analysis

As illustrated in Fig. 16.1, desorption results from strain-induced localization of electronic excitations. The scaling approach to polaronic localization detailed in Section 4.3 is an efficient means of studying polaronic localization. In particular, the scaling parameter of this method directly measures polaronic localization. Here the scaling approach of Section 4.3 is applied to Eq. (16.3) in order to study how electronically induced softening affects self-trapping. The electronically stimulated desorption of adsorbed atoms arising from excitation-induced bond-breaking is treated as a special case of electronically induced softening.

The scaling analysis examines the minimum adiabatic energy, Eq. (16.3), as a function of L, a dimensionless scaling factor which alters the electronic excitation's spatial extent (Emin, 1973a; Emin and Holstein, 1976). In particular, the electronic excitation's spatial coordinate r is replaced by r/L and its normalized wavefunction is replaced by $L^{-d/2}\phi(r/L)$ in Eq. (16.3), where d is the electronic wavefunction's dimensionality. Then, upon introducing a change of variable in the integrations of Eq. (16.3), the minimum adiabatic energy is converted to the energy functional

$$V_{ad}^{min}(L) \equiv \frac{1}{L^2}\left(\frac{\hbar^2}{2m}\int dr |\nabla_r\phi(r)|^2\right) - \frac{1}{L^d}\left(\frac{F^2}{2k}\right)V_c \int dr \frac{|\phi(r)|^4}{\left[1 - \frac{1}{L^d}\left(\frac{k'}{k}\right)V_c|\phi(r)|^2\right]}.$$

(16.4)

Electronically induced softening, having $k' > 0$, lowers the energy functional most severely as L is decreased. Thus, electronically induced softening fosters localization.

For desorption of adsorbed atoms: $d = 2$ and $k' = k$. Then, as illustrated in Fig. 16.2, the energy functional can exhibit two qualitatively distinct behaviors. The energy functional versus L is monotonic when $T_e < V_S$, where:

$$T_e \equiv \frac{\hbar^2}{2m}\int dr |\nabla_r\phi(r)|^2$$

(16.5)

is a measure of the electronic kinetic energy and

$$V_S \equiv \left(\frac{F^2}{2k}\right)V_c \int dr |\phi(r)|^4$$

(16.6)

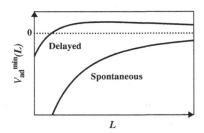

Fig. 16.2 The scaling functional $V_{ad}^{min}(L)$ is plotted versus the scaling length L for delayed desorption $T_e > V_S$ and for spontaneous desorption $T_e < V_S$.

is a measure of the short-range electron–phonon interaction. Alternatively, the energy functional is peaked when $T_e > V_S$.

The first instance, $T_e < V_S$, corresponds to "immediate desorption" since an initially delocalized excitation encounters no impediment to achieving full localization and its inducing atomic desorption (Jennison and Emin, 1983b). The second circumstance, $T_e > V_S$, is termed "delayed desorption" since an initially delocalized excitation must traverse an energy barrier before inducing desorption (Jennison and Emin, 1983b).

The preceding discussion has addressed the role that polaronic localization can play in electronically stimulated desorption on an ideal surface. The surface's reduced dimensionality $d = 2$ and electronically induced softening facilitate the localization required for desorption. As discussed in Section 4.5, surface disorder and traps also generally enhance polaronic localization.

Polaronic phenomena are usually concerned with long-lived carriers associated with atoms' outer shells. By contrast, electronically stimulated desorption generated by bombardment with very high-energy electrons, photons or ions produces holes within atoms' inner shells. Electronically stimulated desorption has been attributed to these holes and the associated Auger-generated multi-holes that co-occupy sites (Feibelman and Knotek, 1978; Madden *et al.*, 1982). Desorption is encouraged by the narrow bandwidths associated with such deep-lying multi-holes (Jennison and Emin, 1983b). However, desorption's efficiency is also limited by the short lifetimes of surface excitations caused by their recombination and diffusion into the solid's bulk.

17

Hopping of light atoms

Important technological problems involve the hopping of light atoms within materials comprising relatively heavy atoms. For example, interstitial diffusion of hydrogen in transition metals is a step in these materials' hydrogen embrittlement. Such hopping is similar to electronic small-polaron hopping in that the masses of the diffusing species are smaller than those of the host. The temperature dependence of a light atom's jump rate is even qualitatively similar to that of an electronic small polaron. Nonetheless, the mass of a hydrogen atom is very much larger than that of an electronic carrier. As will become evident this large mass difference requires generalizing the small-polaronic treatment of hopping to render it applicable to light-atom diffusion. As a result the interpretation of the temperature dependence of a light-atom's jump rate generally differs from that for an electronic small polaron. Most markedly, the resulting light-atom jump rate garners a distinguishing dependence on the atom's isotopic mass.

17.1 High-temperature diffusion: excited coincidences and non-Condon transfers

As described in Chapter 11, a small-polaron's hopping is dependent on the motions of the host material's atoms. At high temperatures these motions become classical. Then a light diffusing particle can only hop when the host atoms' movements bring the energies associated with the diffuser occupying initial and final sites into momentary coincidence.

Increasing the mass of the diffusing particle from that of an electronic carrier to that of a light atom greatly reduces the separations between the energy levels of its bound states. In particular, as illustrated in Fig. 17.1, many energy levels can then be supported within the dual potential wells of a symmetric coincidence configuration. In fact these energy separations will generally be comparable to the thermal energy κT at room temperature. Thus light atoms are generally afforded multiple parallel paths via which they can transfer between sites involved in a coincidence.

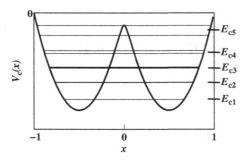

Fig. 17.1 The potential energy for a symmetric two-site coincidence $V_c(x)$ is plotted against a diffusing atom's position x. Even for a very light diffusing atom the separations between successive coincidence energy levels, E_{c1}, E_{c2}, E_{c3}, E_{c4} and E_{c5}, are often small enough for multiple coincidence levels to be thermally accessible at room temperature. The tunnel splitting between each coincidence's two states increases very sharply with coincidence energy. Thus the tunnel splittings in this illustration are only discernable for E_{c4} and E_{c5}.

As described in Sections 11.3–11.6 and depicted in Fig. 11.7 the probability of a diffusing particle availing itself of the transport opportunity offered by the occurrence of a coincidence event depends on the magnitude of the associated transfer energy. For electronic hopping the so-called "Condon approximation", ignoring the dependence of transfer energies on coincidence conditions, is generally invoked. Then the transfer energy for a given electronic carrier is treated as a constant, independent of atomic deformations.

By contrast, the Condon approximation is strongly violated for light-atom diffusion. As illustrated in Fig. 17.1, the transfer energies of light atoms' coincident states, essentially half their tunneling splittings, rise very strongly with coincidence energy. This feature results from a light atom's mass being much larger than that of an electronic carrier. For example, the transfer energy of a particle of mass M tunneling at a coincidence whose energy is E_c below the peak of a barrier of curvature $-k_b$ is $t_{g,g'} \propto \exp[-\pi E_c(M/k_b)^{1/2}/\hbar]$. Thus, the transfer energy's fall with increasing E_c becomes more severe as M is increased: $\partial \ln t_{g,g'}/\partial E_c \sim -\pi(M/k_b)^{1/2}/\hbar$.

Light-atom diffusion is distinguished from electronic hopping by (1) the availability of parallel transfer channels whose (2) transfer energies depend very strongly on coincidence conditions. As a result, unlike the situation for electronic hopping, the predominant hopping of a light atom is not generally associated with the lowest-energy level of the minimum-energy coincidence configuration (Emin *et al.*, 1979). In particular, increasing the mass of the diffuser slows its transfer via low-energy coincidences while increasing the accessibility of higher-energy coincidences with much faster transfer.

Numerical models have been used to study the hopping of hydrogen and its isotopes between shallow interstitial positions in bcc metals (Emin *et al.*, 1979). Transfer energies for low-lying coincidences are found to be so small that transport through them is severely limited. As a result, room-temperature diffusion is dominated by transfer through excited coincidences. The diffusion constant's small activation energy is then just the difference between the energy of the lower of the pair of excited coincidence states and the ground-state energy of an interstitial hydrogen atom. Moreover, excited coincidences' transfer energies are often large enough for the hopping to be adiabatic. In these instances, as described in Chapter 11, an interstitial hydrogen atom hops whenever motions of the transition-metal atoms present it with such excited coincidences. Then the pre-exponential factor of the hydrogen diffusion constant depends on only the motions of the host atoms. In particular, the diffusion constant's pre-exponential factor becomes just $va^2 \approx 10^{-3}$ cm^2/sec, where v is comparable to the Debye temperature of the host metal and a represents the jump distance.

Diffusion through excited coincidences is progressively frozen out as the temperature is reduced. Concomitantly the diffusion constant falls and its temperature dependence gradually weakens thereby producing non-Arrhenius behavior. At lower temperatures the relatively slow residual semiclassical diffusion is associated with only the diffuser's lowest-energy coincidence state.

Early work ignored the possibilities of excited coincidences and breakdowns of the Condon approximation (Flynn and Stoneham, 1970). Thus the diffusion of interstitial atoms in metals was treated like the hopping of electronic small polarons. As depicted in Fig. 11.2, the hopping of an electronic small polaron also displays non-Arrhenius behavior. However, an electronic small polaron's non-Arrhenius behavior occurs well below the host solid's Debye temperature where multi-acoustic-phonon jump processes are frozen out. At very low temperatures in an ideal crystal this freeze-out process culminates with small-polaron hopping occurring primarily via a two-phonon process in which one phonon is absorbed and another phonon of essentially equal energy is emitted (Holstein, 1959b). With the diffuser interacting with acoustic phonons the low-temperature limit of this two-phonon jump rate is proportional to T^7 (Flynn and Stoneham, 1970). This low-temperature behavior only becomes distinguishable from higher-temperature non-Arrhenius behavior when the temperature is lowered well below the host solid's Debye temperature.

17.2 Isotope dependences of light-atoms' high-temperature diffusion constants

Changing isotope significantly alters a light-atom's mass. Deuterium and tritium have double and triple the mass of simple hydrogen, respectively. As such, altering a

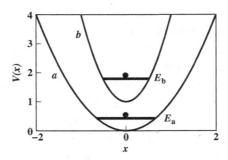

Fig. 17.2 Curves *a* and *b* illustrate the potential energy in the *x*-direction for two configurations of the host material's atoms. Host atoms' movements generally shift both the potential well's minimum and its curvature. The dependence of the diffusing atom's energy, E_a or E_b, on its mass enters through the energy level's dependence on the potential well's curvature.

light-atom's isotope provides a distinctive probe of the light-atom's diffusion process. Indeed, the isotope dependences of light-atoms' diffusion constants can distinguish the mechanism of their hopping from those of electronic small polarons and of classical heavy atoms.

Altering the positions of atoms of the host lattice changes the potential well occupied by the diffusing light atom. As illustrated in Fig. 17.2 modulations of host-atoms' positions generally shift the minimum and the curvature of the potential well containing the diffusing particle. The energy of the light atom changes as its potential well is modified. In particular, Fig. 17.2 shows that the amount that a light atom's energy level is elevated above the potential well's minimum increases as the potential well's curvature increases. The contribution to the total energy shift arising from altering the minimum of the host-atom's potential well is independent of the host-atom's mass. However, the contribution arising from altering the potential well's curvature depends on the diffuser's mass.

Atom–phonon interactions, analogous to electronic polarons' electron–phonon interactions, measure the sensitivities of light-atom's energy levels to shifts of host-atoms' positions. Thus the strengths of these interactions generally fall as a diffusing atom's isotopic mass is increased since heavier isotopes are less sensitive to changes of a potential well's curvature than are lighter isotopes.

The strengths of atom–phonon interactions govern the contribution of host-atom displacements to a light-atom's self-trapping. Since the atom–phonon interaction strength decreases with increasing isotopic mass, the diffusing atom's self-trapping energy also decreases with increasing isotope mass. By itself, this effect would cause heavier isotopes to diffuse more rapidly than lighter isotopes. In particular, the activation energy of an electronic polaron's non-adiabatic semiclassical hopping generally falls as its self-trapping energy falls. In other words, treating a light-atom's hopping as simple non-adiabatic small-polaron hopping implies that the activation energy for the light-atom's hopping will fall as its isotopic mass is increased.

However, as stressed in Section 17.1, the energetic separations between a light-atom's lowest-energy coincidence and its excited coincidences are much less than those for an electronic small polaron. Thus light-atoms' excited coincidences can become thermally accessible even though electronic carriers' excited coincidences are generally ignorable. It is therefore appropriate to consider the diffusion constant for a light atom's thermally activated semiclassical hopping to be a generalization of that for an electronic small polaron:

$$D = va^2 \sum_i P_i \exp\left(-\frac{E_{A,i}}{\kappa T}\right), \tag{17.1}$$

where v is the characteristic vibration frequency of the host atoms and a is the diffusing light-atom's jump distance. Here P_i represents the probability that a diffusing atom hops when presented with the i-th coincidence and $E_{A,i}$ denotes the net energy of that coincidence relative to that of the self-trapped ground-state.

The transfer probability P_i rises strongly with the electronic transfer energy (cf. Fig. 11.7). In addition, as illustrated in Fig. 17.1, a light-atom's excited coincidences have larger transfer energies than those of lower-energy coincidences. Thus relatively large values of P_i tend to be associated with large values of $E_{A,i}$. As a result, when excited coincidences become thermally accessible they tend to dominate a light-atom's high-temperature semiclassical diffusion.

Figure 17.3 illustrates the isotope dependence found from models of high-temperature semiclassical diffusion of interstitial hydrogen in bcc metals (Emin et al., 1979). The highest-temperature diffusion is dominated by coincidences whose transfer energies are large enough to yield adiabatic motion, $P_i \approx 1$. As such, the pre-exponential factor of the diffusion constant is just va^2. The associated activation energy rises as its diffusing isotope's mass increases. This behavior occurs because a larger isotopic mass necessitates forming a higher-energy coincidence to produce a large-enough transfer energy to enable an adiabatic hop.

These features distinguish semiclassical light-atom diffusion from both classical and conventional semiclassical small-polaron hopping. In particular, the diffusion constant's pre-exponential factor being independent of the diffusing light atom's isotopic mass differs from the prediction for classical diffusion. For classical diffusion the diffusion constant's pre-exponential factor is inversely proportional to the square root of the diffuser's mass. In addition, the increase of a diffusion constant's activation energy with its isotope's mass is opposite to the predictions for conventional small-polaron hopping.

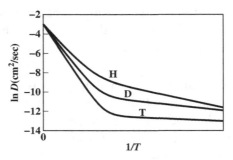

Fig. 17.3 The semiclassical diffusion constants of hydrogen H, deuterium D and tritium T are drawn versus reciprocal temperature in arbitrary units. The three diffusion constants approach their common adiabatic limit $va^2 \approx 10^{-3}$ cm^2/sec at high temperatures. High-temperature diffusion is dominated by adiabatic hopping through excited coincidences. Then the activation energies rise as the isotopic mass is increased. At somewhat lower temperatures diffusion is dominated by non-adiabatic jumps through their lowest-energy coincidences. Then the activation energies fall as the isotopic mass is increased.

At the lowest temperatures of Fig. 17.3 the excited coincidences are frozen out. Then the predominant semiclassical jumps result from their lowest-energy coincidences. As such, a light-atom's semiclassical hop becomes analogous to that of a conventional small polaron. Since the masses of even light atoms are huge relative to those of electronic carriers these hops are generally non-adiabatic. As a result both the diffusion constant's activation energy and its pre-exponential factor decrease with increasing isotopic mass. Figure 17.3 contrasts this low-temperature behavior with the high-temperature behavior where excited coincidences dominate the light atom's diffusion.

17.3 Isotope dependences of light-atoms' low-temperature diffusion constants

Lowering the temperature progressively freezes out excited coincidences and host-atoms' classical motions. At very low temperatures a light-atom's hopping becomes similar to the non-semiclassical hopping of an electronic small polaron. The relatively small masses of electronic carriers compared with those of even light atoms suggests that their transfer energies will generally be small enough to render light-atom's hopping non-adiabatic. Thus a light-atom's low-temperature diffusion becomes formally equivalent to that of a non-adiabatic electronic small-polaron hopping.

As illustrated in panel *a* of Fig. 11.1, low-temperature hopping of an electronic small polaron requires quantum tunneling of both the electronic carrier and the atoms of the host solid. Similarly the low-temperature non-adiabatic small-polaronic

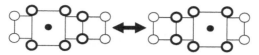

Fig. 17.4 The low-temperature diffusion of a light interstitial atom (solid black dot) occurs as it and surrounding atoms of the host material (open circles) alter their equilibrium positions. For clarity darkened open circles represent those host atoms whose equilibrium positions are shifted in the illustrated jump process.

hopping of a light atom requires tunneling of both the diffusing light atom and the host material's atoms. The initial and final equilibrium locations of the atoms involved in such a tunneling process are illustrated in Fig. 17.4. A light-atom's low-temperature non-adiabatic jump rate is proportional to the absolute squares of both (1) the diffusing atom's transfer energy and (2) the overlap between the wavefunctions for the host atoms' vibrations before and after the hop (Emin, 1975a). Both factors depend on the mass of the diffusing light atom.

Displacements of the equilibrium positions of host atoms surrounding a diffusing light atom depend on its atom–phonon interactions. As discussed in Section 17.2, these interactions measure the sensitivity of the light-atom's energy to changes of the host-atoms' positions. As such, the strengths of these interactions vary inversely with the light-atom's isotopic mass. Thus shifts of the equilibrium positions of host atoms induced by the light-atom's presence decrease upon increasing its isotopic mass. Concomitantly, the depth of the potential well binding the light atom falls as its isotopic mass is increased.

The isotope dependence of a light-atom's low-temperature non-adiabatic small-polaron-like diffusion constant is determined by the isotopic dependences of two factors. The first factor is the absolute square of the overlap between the wavefunctions describing host-atoms' harmonic vibrations about the equilibrium positions they assume in response to a light-atom's hop occupying the hop's initial and final sites. This factor increases as the light-atom's isotopic mass increases because the shifts of host-atoms' equilibrium positions then decrease. The second factor is the absolute square of the light-atom's transfer energy. The depths of the potential wells between which the light atom tunnels decrease as its isotopic mass increases. Nonetheless, the efficacy of the light-atom's tunneling tends to decrease as its isotopic mass increases. All told, the isotope dependence of a light-atom's low-temperature non-adiabatic diffusion is determined by competing effects.

As indicated by the illustration of Fig. 17.3, the isotope dependence of a light-atom's diffusion constant generally changes with temperature. The isotope dependence of a light-atom's diffusion constant may even reverse with falling temperature.

References

Abrikosov, A. A., Gorkov, L. P. and Dzyaloshinski, I. E. (1963). *Methods of Quantum Field Theory in Statistical Physics*, Engelwood Clifts, Prentice-Hall, Sec. 26.1.

Adamowskii, J. (1989). Formation of Fröhlich bipolarons. *Phys. Rev. B* 39, 3649–3652.

Adler, D. (1971). Amorphous semiconductors. In: *CRC Critical Reviews in Solid State Sciences*, Vol. 2, Issue 3, New York, CRC Press, pp. 317–465.

Adler, D., Henisch, H. K. and Mott, N. (1978). The mechanism of threshold switching in amorphous alloys. *Rev. Mod. Phys.* 50, 209–220.

Ambegaokar, V., Halperin, B. I. and Langer, J. S. (1971). Hopping conductivity in disordered systems. *Phys. Rev. B* 4, 2612–2620.

Anderson, P. W. (1950). Antiferromagnetism. Theory of superexchange interaction. *Phys. Rev.* 79, 350–356.

Anderson, P. W. (1958). Absence of diffusion in certain random lattices. *Phys. Rev.* 109, 1492–1505.

Anderson, P. W. (1963). *Concepts in Solids*, New York, Addison-Wesley, pp. 77–95.

Anderson, P. W. (1972). Effect of Franck–Condon displacements on the mobility edge and the energy gap of disordered materials. *Nature Phys. Sci.* 235, 163–165.

Anderson, P. W. (1975). Model for the electronic structure of amorphous semiconductors. *Phys. Rev. Lett.* 34, 953–955.

Anderson, P. W. and Abrahams, E. (1987). Superconductivity theories narrow down. *Nature* 327, 363.

Anderson, P. W. and Hasegawa, H. (1955). Considerations on double exchange. *Phys. Rev.* 100, 675–681.

Aselage, T. L., Emin, D., McCready, S. S. and Duncan, R. V. (1998). Large enhancement of boron carbides' Seebeck coefficients through vibrational softening. *Phys. Rev. Lett.* 81, 2316–2319.

Aselage, T. L., Emin, D. and McCready, S. S. (2001). Conductivities and Seebeck coefficients of boron carbides: "Softening bipolaron" hopping. *Phys. Rev. B* 64, 054302.

Aselage, T. L., Emin, D., McCready, S. S., Venturini, E. L., Rodriguez, M. A., Voigt, J. A. and Headley, T. J. (2003). Metal–semiconductor and magnetic transitions in compensated polycrystalline $La_{1-x}Ca_xMnO_{3-\delta}$ (x = 0.20, 0.25). *Phys. Rev. B.* 68, 134448.

Austin, I. G. and Mott, N. F. (1969). Polarons in crystalline and non-crystalline materials. *Adv. Phys.* 18, 41–102.

Azevedo, L. J., Venturini, E. L., Emin, D. and Wood, C. (1985). Magnetic susceptibility study of boron carbides. *Phys. Rev. B.* 32, 7970–7972.

198

Baily, S. A. and Emin, D. (2006a). Transport properties of amorphous telluride. *Phys. Rev. B* 73, 165211.

Baily, S. A., Emin, D. and Li, H. (2006b). Hall mobility of amorphous $Ge_2Sb_2Te_5$. *Solid State Comm.* 139, 161–164.

Bardeen, J., Cooper, L. N. and Schrieffer, J. R. (1957a). Microscopic theory of superconductivity. *Phys. Rev.* 106, 162–164.

Bardeen, J, Cooper, L. N. and Schrieffer, J. R. (1957b). Theory of superconductivity. *Phys. Rev.* 108, 1175–1204.

Bartram, R. H. and Stoneham, A. M. (1975). On the luminescence and absence of luminescence of *F*-centers. *Solid State Commun.* 17, 1593–1598.

Basani, F., Geddo, M., Iadonisi, G., and Ninno, D. (1991). Variational calculations of bipolaron binding energies. *Phys. Rev. B* 43, 5296–5306.

Batt, R. H., Braun, C. L. and Hornig, J. F. (1968). Electric-field and temperature dependence of photoconductivity. *J. Chem. Phys.* 49, 1967–1968.

Bennett, C. H. (1975). Exact defect calculations in model substances. In: *Diffusion in Solids: Recent Developments*, eds. A. S. Nowick and J. J. Burton, New York, Academic Press, pp. 73–113.

Bennett, C. H. and Alder, B. J. (1968). Persistence of vacancy motion. *Solid State Commun.* 6, 785–789.

Bennett, C. H. and Alder, B. J. (1971). Persistence of vacancy motion in hard sphere crystals. *J. Phys. Chem Solids.* 32, 2111–2122.

Bergeron, E. G. and Emin, D. (1977). Phonon-assisted hopping due to interaction with both acoustical and optical phonons. *Phys. Rev. B.* 15, 3667–3681.

Bishop, S. G., Strom, U. and Taylor, P. C. (1977). Optically induced metastable paramagnetic states in amorphous semiconductors. *Phys. Rev. B* 15, 2278–2294.

Born, M. and Oppenheimer, J. R. (1927). Zur Quantentheorie der Molekeln. *Annalen der Physik* 389, 457–484.

Bösch, M. A., Epworth, R. W. and Emin, D. (1980). Photoluminescence dynamics in chalcogenide glasses and crystals. *J. Non-Crystalline Solids* 40, 587–594.

Bösch, M. A. and Shah, J. (1979). Time-resolved photoluminescence spectroscopy in amorphous As_2S_3. *Phys. Rev. Lett.* 42, 118–121.

Bosman, A. J. and van Daal, H. J. (1970). Small-polaron versus band conduction in some transition-metal oxides. *Adv. Phys.* 19, 1–117.

Callen, H. B. (1960). *Thermodynamics*, New York, John Wiley, Chap. 17.

Chakraverty, B. H., Ranninger, J. and Feinberg, D. (1998). Experimental and theoretical constraints of bipolaronic superconductivity in high-T_c materials: An impossibility. *Phys. Rev. Lett.* 81, 433–436.

Chance, R. R. and Braun, C. L. (1976). Temperature dependence of intrinsic carrier generation in anthracene single crystals. *J. Chem. Phys.* 64, 3573–3581.

Chauvet, O., Emin, D., Forro, L., Zuppiroli, L. and Aselage, T. L. (1996). Spin susceptibility of boron carbides: Dissociation of singlet small-bipolarons. *Phys. Rev. B* 53, 14450–14457.

Clark, A. H. (1967). Electrical and optical properties of amorphous germanium. *Phys. Rev.* 154, 750–757.

Cohen, M. H., Economou, E. N. and Soukoulis, C. M. (1983). Polaron formation near a mobility edge. *Phys. Rev. Lett.* 51, 1202–1205.

Condon, E. (1926). A theory of intensity distribution in band systems. *Phys. Rev.* 28, 1182–1201.

Crevecoeur, C. and de Wit, H. J. (1970). Electrical conductivity of Li doped MnO. *J. Phys. Chem. Solids* 31, 783–791.

De Gennes, P.-G. (1960). Effects of double exchange in magnetic crystals. *Phys. Rev.* 118, 141–154.

de Wit, H. J. (1968). Some numerical calculations on Holstein's small-polaron theory. *Philips Res. Rep.* 23, 449–460.

Dexter, D. L., Klick, C. C. and Russell, G. A. (1955). Criterion for the occurrence of luminescence. *Phys. Rev.* 100, 603–605.

Emin, D. (1970). Correlated small-polaron hopping motion. *Phys. Rev. Lett.* 25, 1751–1755.

Emin, D. (1971a). Vibrational dispersion and small-polaron hopping: Enhanced diffusion. *Phys. Rev. B* 3, 1321–1337.

Emin, D. (1971b). Lattice relaxation and small-polaron hopping motion. *Phys. Rev. B* 4, 3639–3651.

Emin, D. (1971c). The Hall mobility of a small polaron in a square lattice. *Ann. Phys. (N.Y.)* 64, 336–395.

Emin, D. (1972). Energy spectrum of an electron in a periodic deformable lattice. *Phys. Rev. Lett.* 28, 604–607.

Emin, D. (1973a). On the existence of free and self-trapped carriers in insulators: An abrupt temperature-dependent conductivity transition. *Adv. Phys.* 22, 57–116.

Emin, D. (1973b). Aspects of the theory of small-polarons in disordered materials. In: *Electronic and Structural Properties of Amorphous Semiconductor*, eds. P. G. LeComber and J. Mort, London, Academic Press, pp. 261–328.

Emin, D. (1974). Phonon-assisted jump rate in noncrystalline solids. *Phys. Rev. Lett.* 33, 303–307.

Emin, D. (1975a). Phonon-assisted transition rates I: Optical-phonon-assisted hops. *Adv. Phys.* 24, 305–348.

Emin, D. (1975b). Thermoelectric power due to electronic hopping motion. *Phys. Rev. Lett.* 35, 882–885.

Emin, D. (1977a). The sign of the Hall effect in hopping conduction. *Phil. Mag.* 35, 1189–1198.

Emin, D. (1977b). Effect of temperature-dependent shifts on semiconductor transport properties. *Solid State Comm.* 22, 409–411.

Emin, D. (1980a). Electrical and optical properties of amorphous thin films. In: *Amorphous Thin Films and Devices*, ed. L. L. Kazmerski, New York, Academic Press, pp. 17–57.

Emin, D. (1980b). Interactions between small-polaronic particles in solids. In: *The Physics of MOS Insulators*, eds. G. Lucovsky, S. T. Pantelides, and F. L. Galeener, New York, Pergamon Press, pp. 39–43.

Emin, D. (1982). Small Polarons. *Physics Today* 35, 34–40.

Emin, D. (1984). Effect of temperature-dependent energy-level shifts on a semiconductor's Peltier heat. *Phys. Rev. B* 30, 5766–5770.

Emin, D. (1985). Reply to 'The effect of temperature-dependent energies on semiconductor thermopower formulae'. *Phil Mag. B* 51, L53–L56.

Emin, D. (1986). Self-trapping in quasi-one-dimensional solids. *Phys. Rev. B* 33, 3973–3975.

Emin, D. (1987). Today's small polaron. In: *Condensed Matter Physics*, ed. R. L. Orbach, New York, Springer-Verlag, pp. 16–34.

Emin, D. (1990). Theory of electric and thermal transport in boron carbides. In: *The Physics and Chemistry of Carbides, Nitrides and Borides*, ed. R. Freer, London, Kluwer Academic, pp. 691–704.

Emin, D. (1991a). Semiclassical small-polaron hopping in a generalized molecular-crystal model. *Phys. Rev. B* 43, 11720–11724.

Emin, D. (1991b). Additional short-wavelength vibratory modes of a large (bi)polaron. *Phys. Rev. B* 43, 8610–8612.

Emin, D. (1992). Low-temperature ac conductivity of adiabatic small-polaronic hopping in disordered systems. *Phys. Rev. B.* 46, 9419–9427.

Emin, D. (1993). Optical properties of large and small polarons and bipolarons. *Phys. Rev. B* 48, 13691–13702.

Emin, D. (1994a). Phonon-mediated attraction between large bipolarons: Condensation to a liquid. *Phys. Rev. Lett.* 72, 1052–1055.

Emin, D. (1994b). Phonon-mediated attraction between large bipolarons: Condensation to a liquid. *Phys. Rev. B.* 49, 9157–9167.

Emin, D. (1995a). Effect of electronic correlation on the shape of a large bipolaron: Four-lobed planar large bipolaron in an ionic medium. *Phys. Rev. B* 52, 13874–13882.

Emin, D. (1995b). Formation, phase separation and superconductivity of large bipolarons. In: *Polarons and Bipolarons in High-T_c Superconductors and Related Materials*, eds. E. S. H. Salje, A. S. Alexandrov and W. Y. Liang, Cambridge, Cambridge University Press, pp. 80–109.

Emin, D. (1996). Pair breaking in semiclassical singlet small-bipolaron hopping. *Phys. Rev. B* 53, 1260–1268.

Emin, D. (1999). Enhanced Seebeck effect from carrier-induced vibrational softening. *Phys. Rev. B* 59, 6205–6210.

Emin, D. (2000a). Formation and hopping motion of molecular polarons. *Phys. Rev. B* 61, 14543–14553.

Emin, D. (2000b). Singlet-bipolaron formation among degenerate electronic orbitals: Softening bipolarons. *Phys. Rev. B.* 61, 6069–6085.

Emin, D. (2002). Seebeck effect. In: *Wiley Encyclopedia of Electrical and Electronics Engineering Online*, ed. J. G. Webster, New York, John Wiley, pp. 1–44.

Emin, D. (2006a). Unusual properties of icosahedral boron-rich solids. *J. Solid State Chem.* 179, 2791–2798.

Emin, D. (2006b). Current-driven threshold switching of a small polaron semiconductor in a metastable conductor. *Phys. Rev. B* 74, 035206.

Emin, D. (2007). Laser cooling via excitation of localized electrons. *Phys. Rev. B* 76, 024301.

Emin, D. (2008). Generalized adiabatic polaron hopping: Meyer–Neldel compensation and Poole–Frenkel behavior. *Phys. Rev. Lett.* 100, 166602.

Emin, D. (2013). Theory of Meyer–Neldel compensation for adiabatic charge transfer. *Monatsh. Chem.* 144, 3–10.

Emin, D. and Aselage, T. L. (2005). A proposed boron-carbide-based solid-state neutron detector. *J. Appl. Phys.* 97, 013529.

Emin, D., Baskes, M. I. and Wilson, W. D. (1979). Small polaronic diffusion of light interstitials in bcc metals. *Phys. Rev. Lett.* 42, 791–794.

Emin, D. and Bussac, M.-N. (1994). Disorder-induced small-polaron formation. *Phys. Rev. B* 49, 14290–14300.

Emin, D. and Hart, C. F. (1985). Negative differential conductivity in shallow impurity hopping. *Phys. Rev. B* 32, 6503–6509.

Emin, D. and Hart, C. F. (1987a). Existence of Wannier–Stark localization. *Phys. Rev. B.* 36, 7353–7359.

Emin, D. and Hart, C. F. (1987b). Phonon-assisted hopping of an electron on a Wannier–Stark ladder in a strong electric field. *Phys. Rev. B.* 36, 2530–2546.

Emin, D., Hillery, M. S. and Liu, N.-L. H. (1986). Thermally induced abrupt collapse of a shallow donor state in a ferromagnetic semiconductor. *Phys. Rev. B* 35, 2933–2936.

Emin, D., Hillery, M. S. and Liu, N.-L. H. (1987). Thermally induced abrupt shrinking of a donor state in a ferromagnetic semiconductor. *Phys. Rev. B* 35, 641–652.

Emin, D. and Hillery, M. S. (1988). Continuum studies of magnetic polarons and bipolarons in antiferromagnets. *Phys. Rev. B* 37, 4060–4070.

Emin, D. and Hillery, M. S. (1989). Formation of a large singlet bipolaron, application to high-temperature superconductivity. *Phys. Rev. B* 39, 6575–6593.

Emin, D. and Holstein, T. (1969). Studies of small-polaron motion IV: Adiabatic theory of the Hall effect. *Ann. Phys. (N.Y.)* 53, 439–520.

Emin, D. and Holstein, T. (1976). Adiabatic theory of an electron in a deformable continuum. *Phys. Rev. Lett.* 36, 323–326.

Emin, D. and Kriman, A. M. (1986). Transient small-polaron hopping motion. *Phys. Rev. B* 34, 7278–7289.

Emin, D. and Liu, N.-L. H. (1983). Small-polaron hopping in magnetic semiconductors. *Phys. Rev. B* 24, 4788–4798.

Emin, D., Seager, C. H. and Quinn, R. K. (1972). Small polaron hopping conduction in some chalcogenide glasses. *Phys. Rev. Lett.* 28, 813–816.

Emin, D., Ye, J., and Beckel, C. L. (1992). Electron correlation effects in one-dimensional large-bipolaron formation. *Phys. Rev. B* 46, 10710–10720.

Epstein, R. I., Buckwald, M. I., Edwards, B. C., Gosnell, T. R. and Mungan, C. E. (1995). Observation of laser-induced fluorescent cooling of a solid. *Nature London* 377, 500–502.

Feibelman, P. J. and Knotek, M. L. (1978). Reinterpretation of electron-stimulated desorption data from chemisorption systems. *Phys. Rev. B*, 18, 6531–6539.

Feynman, R. P. (1972). *Statistical Mechanics: A Set of Lectures*, New York, Addison-Wesley, Sec. 2.6; Chap. 11.

Fischer, R., Heim, U., Stern, F., and Weiser, K. (1971). Photoluminescence of amorphous $2As_2Te_3$-As_2Se_3 films. *Phys. Rev. Lett.* 26, 1182–1185.

Flynn, C. P. and Stoneham, A. M. (1970). Quantum theory of diffusion with application to light interstitials in metals. *Phys. Rev. B* 1, 3966–3978.

Fork, R. L., Shank, C. V., Glass, A. M., Migus, A., Bösch, M. A. and Shah, J. (1979). Picosecond dynamics of optically induced absorption in the bandgap of As_2S_3. *Phys. Rev. Lett.* 43, 394–398.

Forro, L., Chauvet, O., Emin, D., Zuppiroli, L., Berger, H. and Levy, F. (1994). High mobility *n*-type charge carriers in large single crystals of anatase (TiO_2). *J. Appl. Phys.* 75, 633–635.

Franck, J. (1926). Elementary processes of photochemical reactions. *Trans. Faraday Soc.* 21, 536–542.

Frenkel, J. (1938). On pre-breakdown phenomena in insulators and electronic semiconductors. *Phys. Rev.* 54, 647–648.

Friedman, L. (1963). Hall effect in the polaron-band regime. *Phys. Rev.* 131, 2445–2456.

Friedman, L. (1964). Transport properties of organic semiconductors. *Phys. Rev.* 133, A1668–A1679.

Friedman, L. and Holstein, T. (1963). Studies of polaron motion Part III: The Hall mobility of a small polaron. *Ann. Phys. (N.Y.)* 21, 494–549.

Fritzsche, H. (2006). Why chalcogenides are ideal materials for Ovshinsky's Ovonic threshold and memory devices. *Phys. Chem. Glasses* 47, 77–82.

Fröhlich, H. (1963). Introduction to the theory of the polaron. In: *Polarons and Excitons*, ed. C. G. Kuper and G. D. Whitfield, New York, Plenum Press, pp. 1–32.

Fröhlich, H. and Sewell, G. L. (1959). Electric conduction in semiconductors. *Proc. Phys. Soc. (London)* 74, 643–647.

Geballe, T., (1959). Group IV semiconductors. In: *Semiconductors*, ed. N. B. Hannay, New York, Reinhold, pp. 313–388.

Ghosh, P. K. and Spear, W. E. (1968). Electronic transport in liquid and solid sulphur. *J. Phys. C* 1, 1347–1358.

Gibbons, D. J. and Spear, W. E. (1966). Electron hopping transport and trapping phenomena in orthorhombic sulphur crystals. *J. Phys. Chem. Solids*, 27, 1917–1925.

Godart, C., Mauger, A., Desfours, J. P. and Achard, J. C. (1980). Physical properties of EuO versus electronic concentrations. *J. de Physique* 41, C5-205–C5-214.

Grant, A. J., Moustakas, T. D., Penny, T. and Weiser, K. (1974). Conduction in localized band-tail and in extended states. I. Experimental studies of transport in amorphous arsenic telluride. In: *Amorphous and Liquid Semiconductors*, eds. J. Stuke and W. Brenig, London, Taylor & Francis, pp. 325–333.

Heikes, R. R., Maradudin, A. A. and Miller, R. C. (1963). Une etude des proprietes de transport des semiconducteurs de valence mixte. *Ann. Phys. (Paris)* 8, 733–746.

Heikes, R. R. and Ure, R. W. (1961). *Thermoelecticity: Science and Engineering*, New York, Interscience.

Henry, C. H. and Lang, D. V. (1977). Nonradiative capture and recombination by multi-phonon emission in GaAs and GaP. *Phys Rev. B* 15, 989–1016.

Herring, C. (1961). The current state of transport theory. In: *Proceedings of the International Conference on Semiconductor Physics*, Prague 1960, New York, Academic Press, pp. 60–67.

Hillery, M. S., Emin, D. and Liu, N.-L. H. (1988). Effect of an applied magnetic field on the abrupt donor collapse in a ferromagnetic semiconductor. *Phys. Rev. B* 38, 9771–9777.

Hiramoto, H. and Toyozawa, Y. (1985). Inter-polaron interaction and bipolaron formation I. *J. Phys. Soc. Jap.* 54, 245–259.

Holstein, T. (1959a). Studies of polaron motion Part 1: The molecular-crystal model. *Ann. Phys. (N.Y.)* 8, 325–342.

Holstein, T. (1959b). Studies of polaron motion Part II: The "small" polaron. *Ann. Phys. (N.Y.)* 8, 343–389.

Holstein, T. (1973). Sign of the Hall coefficient in hopping-type charge transport. *Phil. Mag.* 27, 225–233.

Holstein, T. (1981). Dynamics of self-localized charge carriers in quasi 1-D solids. *Mol. Cryst. Liq. Cryst.* 77, 235–252.

Holstein, T. and Friedman, L. (1968). Hall mobility of the small polaron. II. *Phys. Rev.* 165, 1019–1031.

Hughes, R. C. (1971a). Geminate recombination of x-ray excited electron–hole pairs in anthracene. *J. Chem. Phys.* 55, 5442–5447.

Hughes, R. C. (1971b). Geminate recombination of x-ray excited carriers in organic solids: Poly-*n*-vinylcarbazole. *Chem. Phys. Lett.* 8, 403–406.

Hundley, M. F. and Neumeier, J. J. (1997). Thermoelectric power of $La_{1-x}Ca_xMnO_{3+\delta}$: Inadequacy of the nominal $Mn^{3+/4+}$ valence approach. *Phys. Rev. B* 55, 11511–11515.

Ing, Jr, S. W., Neyhart, J. H. and Schmidlin, F. (1971). Charge transport and photoconductivity in amorphous arsenic trisulfide films. *J. Appl. Phys.* 42, 696–703.

Jahn, H. A. and Teller, E. (1937). Stability of polyatomic molecules in degenerate electronic states. I. Orbital degeneracy. *Proc. Roy. Soc. A*, 161, 220–235.

Jaime, M., Hardner, H. T., Salamon, M. B., Rubinstein, M., Dorsey, P. and Emin, D. (1997). Hall effect sign anomaly and small-polaronic conduction in $(La_{1-x}Gd_x)_{0.67}Ca_{0.33}MnO_3$. *Phys. Rev. Lett.* 78, 951–954.

Jennison, D. R. and Emin, D. (1983a). Localization of surface excitations and stimulated desorption. *J. Vac. Sci. Technol. A* 1, 1154–1156.

Jennison, D. R. and Emin, D. (1983b). Strain-induced localization and electronically stimulated desorption and dissociation. *Phys. Rev. Lett.* 51, 1390–1393.

Joannopolous, J. D. and Cohen, M. H. (1973). Electronic properties of complex crystalline and amorphous phases of Ge and Si. I: Density of states and band structures. *Phys. Rev. B* 7, 2644–2657.

Kane, E. O. (1959). Zener tunneling in semiconductors. *J. Phys. Chem. Solids*, 12, 181–188.

Karplus, R. and Luttinger, J. M. (1954). Hall effect in ferromagnets. *Phys. Rev.* 95, 1154–1160.

Kastner, M., Adler, D. and Fritzsche, H. (1976). Valence-alternation model for localized gap states in lone-pair semiconductors. *Phys. Rev. Lett.* 37, 1504–1507.

Kasuya, T., Yanase, A. and Takeda, T. (1970). Stability condition for the paramagnetic polaron in a magnetic semiconductor. *Solid State Commun.* 8, 1543–1546.

Kepler, R. G. and Coppage, F. N. (1966). Generation and recombination of holes and electrons in anthracene. *Phys. Rev.* 151, 610–614.

Kertesz, M., Riess, I., Tannhauser, D. S., Langpape, R. and Rohr, F. J. (1982). Structure and electrical conductivity of $La_{0.84}Sr_{0.16}MnO_3$. *J. Solid State Chem.* 42, 125–129.

Klaffe, G. R. and Wood, C. (1976). The Hall effect in amorphous As_2Se_3. In: *Physics of Semiconductors: Proceedings of the Thirteenth International Conference*, ed. F. G. Fumi, London, North-Holland, pp. 545–548.

Klinger, M. I. (1963). Quantum theory of non-steady-state conductivity in low mobility solids. *Phys. Lett.* 7, 102–104.

Kolodiazhnyi, T. and Wimbush, S. C. (2006). Spin singlet small bipolarons in Nb-doped $BaTiO_3$. *Phys. Rev. Lett.* 96, 246404.

Kramers, H. A. (1934). L'interaction entre les atomes magnétogènes dans un cristal paramagnétique. *Physica* 1, 182–192.

Kübler, J. and Vigren, D. T. (1975). Magnetically controlled electron localization in Eu-rich EuO. *Phys. Rev. B* 11, 4440–4449.

Lakkis, S., Schlenker, C., Chakraverty, B. K., Buder, R. and Marezio, R. (1976). Metal–insulator transitions in Ti_4O_7 single crystals: Crystal characterization, specific heat, and electron paramagnetic resonance. *Phys. Rev. B.* 14, 1429–1440.

Landau, L. (1933). On the motion of electrons in a crystal lattice. *Phys. Z. Sowjetunion* 3, 664–665.

Landau, L. (1946). On the thermodynamics of photoluminescence. *J. Phys. (Moscow)* 10, 503–506.

Landau, L. D. and Lifshitz, E. M. (1958). *Statistical Physics*, London, Pergamon Press, Secs. 67, 73.

LeComber, P. G., Jones, D. and Spear, W. E. (1977). Hall effect and impurity conduction in substitutionally doped amorphous silicon. *Phil. Mag.* 35, 1173–1187.

Leroux-Hugon, P. (1972). Dielectric constant of an exchange-polarized electron gas and the metal–insulator transition in EuO. *Phys. Rev. Lett.* 29, 939–943.

Lewis, A. J. (1976a). Conductivity and thermoelectric power of amorphous germanium and amorphous silicon. *Phys. Rev. B* 13, 2565–2575.

Lewis, A. J. (1976b). Use of hydrogenation in the study of the transport properties of amorphous germanium. *Phys. Rev. B* 14, 658–668.

Lifshitz, E. M. and Pitaevskii, L. P. (1980). *Statistical Physics Part 2: Theory of the Condensed State*, Oxford, Pergamon Press, Secs. 34, 44.

Liu, N.-L. H. and Emin, D. (1979). Double exchange and small-polaron hopping in magnetic semiconductors. *Phys. Rev. Lett.* 42, 71–74.

Liu, N.-L. H. and Emin, D. (1984). Thermoelectric power of small polarons in magnetic semiconductors. *Phys. Rev. B.* 30, 3250–3256.

Low, G. G. (1963). Application of spin wave theory to three magnetic salts. *Proc. Phys. Soc (London)* 82, 992–1001, Fig. 1.

Madden, H. H., Jennison, D. R., Traum, M. M., Margaritondo, G. and Stoffel, N. G. (1982). Correlation of H^+-desorption threshold with localized state observed in Auger line shape Si(001):H. *Phys. Rev. B* 26, 896–902.

Mauger, A. (1983). Magnetic polaron: Theory and Experiment. *Phys. Rev. B* 27, 2308–2324.

Marcus, R. A. (1960). Exchange reactions and electron transfer reactions including isotopic exchange. *Discuss. Faraday Soc.* 29, 21–31.

Mauger, A. and Mills, D. L. (1985). Role of conduction-electron-local moment exchange in antiferromagnetic semiconductors: Ferrons and bound magnetic polarons. *Phys. Rev. B* 31, 8024–8033.

Melz, P. (1972). Photogeneration of trinitrofluorenone-poly(N-vinylcarbazole). *J. Chem. Phys.* 57, 1694–1699.

Mendez, E. E., Agulló-Rueda, F. and Hong, J. M. (1988). Stark localization in GaAs-GaAlAs superlattices under an electric field. *Phys. Rev. Lett.* 60, 2426–2429.

Methfessel, S. and Mattis, D. C. (1968). Magnetic semiconductors. *Handbuch der Physik* 18/1, 389–562.

Meyer, Von W. and Neldel, H. (1937). Über die beziehungen zwischen der energiekonstanten E und der mengenkonstanten a in der leitwerts-temperaturformel bei oxydischen halbleitern. *Z. Tech. Phys.* 18 No. 12, 588–593.

Miller, A. and Abrahams, E. (1960). Impurity conduction at low concentrations. *Phys. Rev.* 120, 745–755.

Miller, R. C., Heikes, R. R., and Mazelsky, R. (1961). Model for the electronic transport properties of mixed valency semiconductors. *J. Appl. Phys.* 32, 2202–2206.

Moizhes, B. Ya. and Suprun, S. G. (1984). Bipolarons in *n*-type barium titanate(?). *Sov. Phys. Solid State* 26, 544–545.

Mott, N. F. (1974). *Metal–Insulator Transitions*, London, Taylor & Francis.

Mott, N. F. and Davis, E. A. (1979). *Electronic Processes in Non-Crystalline Materials*, second edition. Oxford, Clarendon Press.

Mott, N. F. and Stoneham, A. M. (1977). The lifetime of electrons, holes and excitons before self-trapping. *J. Phys. C: Solid State Phys.* 10, 3391–3398.

Moustakas, T. D. and Paul, W. (1977). Transport and recombination in sputtered hydrogenated amorphous germanium. *Phys. Rev. B* 16, 1564–1576.

Mytilineou, E. and Davis, E. A. (1977). Thermopower, conductivity and Hall effect in amorphous arsenic. In: *Amorphous and Liquid Semiconductors*, ed. W. E. Spear, Edinburgh, University of Edinburgh Press, pp. 632–636.

Nagaev, E. L. (1967). Ground state and anomalous magnetic moment of conduction electrons in an antiferromagnetic semiconductor. *Sov. Phys. JETP Lett.* 6, 18–20.

Nagaev, E. L. (1976). Ferromagnetic and antiferromagnetic semiconductors. *Sov. Phys. Usp.* 18, 863–892.

Nagels, P. (1980). Experimental Hall effect data for a small-polaron semiconductor. In: *The Hall Effect and its Applications*, eds. C. L. Chien and C. R. Westgate, New York, Plenum Press, pp. 253–280.

Nagels, P., Callaerts, R. and Denayer, M. (1974). Conduction in extended and localized states in the amorphous system As-Te-Si. In: *Amorphous and Liquid Semiconductors*, eds. J. Stuke and W. Brenig, London, Taylor & Francis, pp. 867–876.

Oliver, M. R., Kafalas, J. A., Dimmock, J. O. and Reed, T. B. (1970). Pressure dependence of the electrical resistivity of EuO. *Phys, Rev. Lett.* 24, 1067–1067.

Oliver, M. R., Dimmock, J. O., McWhorter, A. L. and Reed, T. B. (1972). Conductivity studies of europium oxide. *Phys. Rev. B* 5, 1078–1098.

Onsager, L. (1934). Deviations of Ohm's law in weak electrolytes. *J. Chem. Phys.* 2, 599–615.

Onsager, L. (1938). Initial recombination of ions. *Phys. Rev.* 54, 554–557.

Öpik, U. and Pryce, M. H. L. (1957). Studies of the Jahn–Teller effect I: A survey of the static problem. *Proc. Roy. Soc. London A* 238, 425–447.

Pai, D. M. and Enck, R. C. (1975). Onsager mechanism of photogeneration in amorphous selenium. *Phys. Rev. B* 11, 5163–5174.

Palstra, T. T. M., Raminez, A. P., Cheong, S.-W., Zegarski, B. R., Schiffer, P. and Zaanen, J. (1997). Transport mechanisms in doped $LaMnO_3$: Evidence for polaron formation. *Phys. Rev. B* 56, 5104–5107.

Paul, W. and Connell, G. A. N. (1976). Structural modelling of disordered semiconductors. In: *Physics of Structurally Disordered Solids*, ed. S. S. Mitra, New York, Plenum, pp. 45–91.

Pekar, S. I. (1963). *Investigations of the Electronic Theory of Crystals*, Washington: U.S. Government Printing Office, US AEC Report-tr-5575 [Russian original: Gostekhizdat, Moscow, 1951).

Penny, T., Shafer, M. W. and Torrance, J. B. (1972). Insulator–metal transition and long-range magnetic order in EuO. *Phys. Rev. B* 5, 3669–3674.

Penrose, O. and Onsager, L. (1956). Bose–Einstein condensation and liquid helium. *Phys. Rev.* 104, 576–584.

Petrich, G., von Molnar, S. and Penney, T. (1971). Exchange-induced autoionization in Eu-rich EuO. *Phys. Rev. Lett.* 26, 885–888.

Pincus, P. (1972). Polaron effects in the nearly atomic limit of the Hubbard model. *Solid State Commun.* 11, 51–54.

Pitaevskii, L. P. (1959). Properties of the spectrum of elementary excitations near the disintegration threshold of the excitations. *Sov. Phys. JETP* 9, 830–837.

Pollak, M. and Geballe, T. H. (1961). Low-frequency conductivity due to hopping processes in silicon. *Phys. Rev.* 122, 1742–1753.

Pollak, M. and Pike, G. E. (1972). AC conductivity of glasses. *Phys. Rev. Lett.* 28, 1449–1451.

Pooley, D. (1966a). F-center production in alkali halides by electron–hole recombination and a subsequent [110] replacement sequence: A discussion of the electron–hole recombination. *Proc. Phys. Soc.* 87, 245–256.

Pooley, D. (1966b). [110] anion replacement sequences in alkali halides and their relation to F-centre production by electron–hole recombination. *Proc. Phys. Soc.* 87, 257–262.

Pringsheim, P. (1929). Zwei Bemerkungen über den Unterschied von Lumineszenz und Temperaturstrahlung. *Z. Phys.* 57, 739–746. Two remarks on the difference of luminescence and thermal radiation.

Rashba, E. E. (1957). Theory of strong interactions of electron excitations with lattice vibrations in molecular crystals. *Opt. Spektrosk.* 2, 75–87.

Reik, H. G. and Heese, D. (1967). Frequency dependence of the electrical conductivity of small polarons for high and low temperatures. *J. Phys. Chem. Solids* 28, 581–596.

Robertson, J. (1975). Valence *s* bands in chalcogens and amorphous arsenic. *J. Phys. C.* 8, 3131–3136.

Samara, G. A., Tardy, H. L., Venturini, E. L., Aselage, T. L. and Emin, D. (1993). AC hopping conductivities, dielectric constants and reflectivities of boron carbides. *Phys. Rev. B* 48, 1468–1477.

Schafroth, M. R. (1955). Superconductivity of a charged ideal Bose gas. *Phys. Rev.* 100, 463–475.

Schein, L. B. and Borsenberger, P. M. (1993). Hole mobilities in hydrazone-doped polycarbonate and poly(styrene). *Chem. Phys.* 177, 773–781.

Schirmer, O. F., Imlau, M., Merschjann, C. and Schoke, B. (2009). Electron small polarons and bipolarons in LiNbO₃. *J. Phys. Condens. Matter* 21, 12301.

Schirmer, O. F. and Salje, E. (1980). Conduction bipolarons in low-temperature crystalline WO3-x. *J. Phys. C: Solid State Phys.* 13, L1067–L1072.

Schirmer, O. F. and von der Linde, D. (1978). Two-photon- and X-ray-induced Nb^{4+} and O⁻ small polarons in LiNbO₃. *Appl. Phys. Lett.* 33, 35–38.

Schultz, T. D. (1963). Feynman's path-integral method applied to the equilibrium properties of polarons and related problems. In: *Polarons and Excitons*, eds. C. G. Kuper and G. D. Whitfield, New York, Plenum, pp. 71–121; p. 110.

Schüttler, H.-B. and Holstein, T. (1986). Dynamics and transport of a large acoustic polaron in one dimension. *Ann. Phys. (N.Y.)* 166, 93–163.

Seager, C. H. (1971). Electronic Hall mobility in the alkaline-earth fluorides. *Phys. Rev. B* 3, 3479–3484.

Seager, C. H. and Emin, D. (1970). High-temperature measurements of the electron Hall mobility in the alkali halides. *Phys. Rev. B* 2, 3421–3431.

Seager, C. H., Emin, D. and Quinn, R. K. (1973). Electrical transport and structural properties of bulk As-Te-I, As-Te-Ge and As-Te chalcogenide glasses. *Phys. Rev. B* 8, 4746–4760.

Seager, C. H., Knotek, M. L. and Clark, A. H. (1974). DC transport properties of evaporated a-Ge films as a function of annealing. In: *Amorphous and Liquid Semiconductors*, eds. J. Stuke and W. Brenig, London, Taylor & Francis, pp. 1133–1138.

Seager, C. H. and Quinn, R. K. (1975). DC electronic transport in binary arsenic chalcogenide glasses. *J. Non-Cryst. Solids* 17, 386–400.

Shah, J. and Bösch, M. A. (1979). Band-to-band luminescence in amorphous solids: Implications for the nature of electronic band states. *Phys. Rev. Lett.* 42, 1420–1423.

Shao, Y., Hughes, R. A., Dabkowski, A., Radtke, G., Gong, W. H., Preston, J. S. and Botton, G. A. (2008). Structural and transport properties of epitaxial niobium-doped BaTiO₃ films. *Appl. Phys. Lett.* 93, 192114.

Sil, A., Giri, A. K. and Chatterjee, A. (1991). Stability of the Fröhlich bipolaron in two and three dimensions. *Phys. Rev. B* 43, 12642–12645.

Street, R. A. and Mott, N. F. (1975). States in the gap in glassy semiconductors. *Phys. Rev. Lett.* 35, 1293–1296.

Suprun, S. G. and Moizhes, B. Ya. (1982). Role of electron correlation in the creation of a Pekar bipolaron. *Sov. Phys. Solid State* 24, 903–905.

Tanaka, J., Umehara, M., Tamura, S., Tsukioka, M., and Ehara, S. (1982). Study on electric resistivity and thermoelectric power in (La$_{0.8}$Ca$_{0.2}$)MnO$_{3-y}$. *J. Phys. Soc. Jpn.* 51, 1236–1242.

Tanaka, J., Nozaki, H., Horiuchi, S. and Tsukioka, M. (1983a). Etude expérimentale du comportement magnétique du (La$_{0.8}$Ca$_{0.2}$)MnO₃ préparé par la méthode de co-précipitation. *J. Physique-Lett.* 44, L129–L134.

Tanaka, J., Takahashi, K., Yukino, K. and Horiuchi, S. (1983b). Electrical conduction of (La$_{0.8}$Ca$_{0.2}$)MnO₃ with homogeneous ionic distribution. *Phys. Stat. Solidi* 80, 621–630.

Thorpe, M. F., Weaire, D., and Alben, R. (1973). Electronic properties of an amorphous solid. III. The cohesive energy and the density of states. *Phys. Rev. B* 7, 3777–3788.

Torrence, J. B., Shafer, M. W. and McGuire, T. R. (1972). Bound magnetic polarons and the insulator–metal transition in EuO. *Phys. Rev. Lett.* 29, 1168–1171.

Toyozawa, Y. (1961). Self-trapping of an electron by the acoustical mode of lattice vibration. I. *Prog. Theor. Phys.* 26, 29–44.

Umehara, M. (1981). Effect of the electron–phonon interaction on the self-trapped magnetic polaron. *J. Phys. Soc. Jpn.* 50, 1082–1090.

Umehara, M. and Kasuya, T. (1972). A theory for a self-trapped antiferromagnetic polaron at $T = 0$ K. *J. Phys. Soc. Jpn.* 33, 602–615.

Verbist, G., Peters, F. M. and Devreese, J. T. (1991). Large bipolarons in two and three dimensions. *Phys. Rev. B* 43, 2712–2720.

Vinetskii, V. L. (1961). Bipolar states of current carriers in ionic crystals. *Sov. Phys. JETP* 13, 1023–1028.

Vinetskii, V. L. and Giterman, M. Sh. (1958). On the theory of the interaction of "excess" charges in ionic crystals. *Sov. Phys. JETP* 6, 560–564.

Vitins, J. and Wachter, P. (1975). Doped EuTe: A mixed magnetic system. *Phys. Rev. B* 12, 3829–3839.

Volger, J. (1954). Further experimental investigations on some ferromagnetic oxidic compounds of manganese with perovskite structure. *Physica XX*, 49–66.

von Mühlenen, A., Errien, N., Schaer, M., Bussac, M.-N. and Zuppiroli, L. (2007). Thermopower measurements on pentacene transistors. *Phys. Rev. B* 75, 115338.

Wannier, G. H. (1959). *Elements of Solid State Theory*, London, Cambridge University Press, pp. 190–193.

Wood, C. (1986). Transport properties of boron carbides. In: *Boron-Rich Solids*, eds. D. Emin, T. L. Aselage, C. L. Beckel, I. A. Howard and C. Wood, New York, American Institute of Physics, pp. 206–215.

Yamashita, J. and Kurosawa, T. (1958). On electronic current in NiO. *J. Phys. Chem. Solids* 5, 34–43.

Zener, C. (1932). Non-adiabatic crossing of energy levels. *Proc. Roy. Soc.* A137, 696–702.

Index